THE
VELVET CLAW

A NATURAL HISTORY OF THE CARNIVORES

THE
VELVET CLAW

A NATURAL HISTORY OF THE CARNIVORES

David
Macdonald

BBC BOOKS

Picture credits

Ardea pages 29, 39 (Joanna van Gruisen), 66 (Y. Arthus-Bertrand), 93 (John Daniels), 100 (Clem Haagner), 102 (Stefan Meyers), 119 (Clem Haagner), 168–9 (S. Roberts), 188–9 (Jack A. Bailey), 220 (Joanna van Gruisen) and 229 (Clem Haagner); **Stuart Brooks Animation** pages 12–13, 16–17, 18, 19, 20–21, 25, 52 (*top three*), 52–3, 54, 55, 56, 79, 82, 83, 96–7, 120, 127, 129 (*top*), 130, 148–9, 161, 162, 183 and 184–5; **Bruce Coleman** pages 30 (Carol Hughes), 31 (Rod Williams), 45 (Günter Ziesler), 151 (Bob & Clara Calhoun), 152 (Hans Reinhard), 154 (Hans Reinhard), 157 (Rod Williams), 175 (Gerald Cubitt), 181 (Wayne Lankinen), 206–7 (Kim Taylor) and 212 (Erwin & Peggy Bauer); **DRK** page 46 (Stanley Breeden); **4:2:2 Videographic Design** pages 22, 24, 27, 52 (*bottom*), 53, 58, 59, 63, 125 and 129 (*bottom*); **Steve Kaufman** page 89; **Richard Matthews** pages 142–3; **Gus Mills** pages 137 and 139; **NHPA** pages 70 (Gérard Lacz), 2 and 73 (Anthony Bannister), 84 (Orion Press), 112–13 (Martin Wendler), 132 (Anthony Bannister), 160 (Andy Rouse), 200–1 (Stephen Krasemann), 204 (Anthony Bannister), 223 (Anthony Bannister) and 245 (Peter Johnson); **Oxford Scientific Films** pages 32 (Animals Animals/Mickey Gibson), 35 (Doug Allan), 48–9 (Animals Animals/Marty Stouffer Productions), 61 (Owen Newman), 67 (Rafi Ben-Shahar), 81 (Partridge Films/ Jim Clare), 85 (Konrad Wothe), 115 (Tom Ulrich), 153 (Animals Animals/C. C. Lockwood), 186 (Tom Ulrich), 194–5 (Lon Lauber), 226–7 (David Macdonald), 231 (Animals Animals/E. R. Degginger), 234 (David Macdonald) and 240–1 (David Macdonald); **Planet Earth Pictures** pages 65 (Richard Matthews), 99 (Jonathan Scott), 101 (Jim Brandenburg) and 109 (Richard Matthews); **Philip Richardson** page 124; **Claudio Sillero** page 105; **Survival Anglia** pages 57 (Bruce Davidson), 122 (Alan Root), 156 (Alan Root) and 198 (Jeff Foott).

This book is published to accompany the
television series entitled *The Velvet Claw*
which was first broadcast in Autumn 1992
Published by BBC Books,
a division of BBC Enterprises Limited,
Woodlands, 80 Wood Lane
London W12 0TT

First Published 1992
© David Macdonald 1992
ISBN 0 563 20844 9

Designed by Grahame Dudley Associates
Family trees designed by Eugene Fleury
Set in 11/14 pt Century Old Style
Printed and bound in Great Britain by Butler & Tanner Ltd, Frome and London
Colour separation by Technik Ltd, Berkhamsted
Jacket printed by Lawrence Allen Ltd, Weston-super-Mare

ALL THE ANIMATIONS FEATURED IN *THE VELVET CLAW*
WERE SPECIALLY CREATED FOR THE BBC BY STUART BROOKS
ANIMATION, LONDON.
THE WIRE FRAMES WERE SPECIALLY CREATED FOR THE
BBC BY 4:2:2 VIDEOGRAPHIC DESIGN, BRISTOL.

Contents

For
Jenny, Ewan, Fiona and Isobel

Preface & Acknowledgements

Science is a collaborative activity. Each researcher's studies are fuelled by facts and figures gathered by a host of colleagues, and ideas often take shape in many minds independently. Therefore, in writing this book and the seven-part BBC television series it accompanies, I have been supported by the labours and ideas of literally hundreds of colleagues. In more technical publications biologists conventionally acknowledge this debt by citing the names of colleagues to whom each discovery is attributed. However, in a book for the general reader the consensus is that peppering the text with names would be distracting. Instead, I cannot emphasise too strongly at the outset that the following pages are based on the labours of a generation of Carnivore biologists.

It would consume pages to name all those whose work I have cited. However, the following lists those whose discoveries have particularly enriched the text. I can only hope I have represented their work correctly.

D. Adams, William Akersten, Roland Albignac, Peter Apps, Ted Bailey, Melissa Bateson, Mark Bekoff, Brian Bertram, Don Bowen, C.K. Brain, G. Brandes, David Bruce, David Bygott, Lu Carbyn, Tim Caro, Colin Clark, Tom Clarke, Anthony Collins, Laurie Corbett, Orin Courtenay, Scott & Nancy Creel, Tamar Dayan, Seán Doolan, Nigel Dunstone, Nichole Duplaix, Marion East, Sam Erlinge, Paul Errington, James Estes, Griff Ewer, John Fanshaw, Clare Fitzgibbon, George & Lory Frame, Laurence Frank, Eduardo Fuentes, Eli Geffen, Dennis Gilbert, Darren Gills, Michael Gilpin, Luc-Alain Giraldeau, John Gittleman, J. Gliwicz, Matthew Gompper, Martyn Gorman, Richard Goss, Chris Gow, Manfred Gutjahr, Ruth Harkness, David Henry, Pall Hersteinsson, Ray Hewson, Heribert Hofer, Maurice Hornocker, Giora Ilany, Fabian Jaksic, Susan Jenks, Charles Jonkel, Gillian Kerby, Kim King, Hans Kruuk, Karl Kugelschafter, Hugo van Lawick, Niles Lehman, Richard Leonard, Paul Leyhausen, Erik Lindstrom, Nick Lunn, Kathy Lyons, Barbara Maas, J. MacLean, Audrey Magoun, Larry Marshall, Dave Mech, John Messick, Gus Mills, Patricia Moehlman, Jan Nel, Ralph Nelson, Alan Newsome, A.J. Nicholson, Steve O'Brien, Peter Oeltgen, Nikita Ovsyanikov, Mark & Dehlia Owens, Jane Packard, Craig Packer, Warner Passanisi, Rolf Petersen, Roger Powell, Anne Pusey, Leonard Radinsky, Anne Rasa, Philip Richardson, Lynn Rodgers, Jon Rood, James Russel, Bob Savage, George Schaller, Torbjorn von Schantz,

Claudio Sillero, Michael Soule, Ian Stirling, Frederick Szalay, Mitch Taylor, Blaire Van Valkenburgh, Denis Voigt, Peter Waser, Robert Wayne, Lars Werdelin, Chris Wozencraft and Yoram Yom-Tov.

Several friends have been kind enough to comment on a draft of the whole book. As world leaders in the field, I am profoundly grateful for their efforts to keep me on the straight and narrow, and record my heartfelt thanks to Tamar Dayan, Joshua Ginsberg, John Gittleman, Gus Mills, Bob Savage, Blaire Van Valkenburgh, Robert Wayne, and Lars Werdelin. Others have generously criticised particular sections, often contributing their own unpublished observations: Tim Caro, Scott Creel, Laurence Frank, Kathy Lyons, James Malcolm, Roger Powell, Anne Rasa, Philip Richardson, Mikael Sandell, Claudio Sillero and Ian Stirling.

In preparing the book I was greatly aided by members of Oxford University's Wildlife Conservation Research Unit: Laura Handoca, Clare Hawkins, Paul Stewart and Rosie Woodroffe worked tirelessly to help me unearth interesting stories. The final product also owes much to the editorial skill of Judy Maxwell. This book and *The Velvet Claw* television series have advanced side by side and I thank the BBC producers Melinda Barker, Andrew Jackson, Chris McFarling, Paul Reddish and Paul Stewart for their support and loyalty. I also thank the executive producer, Mike Beynon, for inviting me to join this team.

This book will be published on the twentieth anniversary of my professional involvement with Carnivores. They have led me from Borneo to Brazil and from South Africa to Iceland, and around the globe I have relished the friendship of naturalists who shared the thrill of these enthralling creatures. Their collected wisdom has fashioned my ideas and enhanced any virtue this book may have. Since 1977 my research group at Oxford has been home to a succession of Carnivore enthusiasts: Peter Apps, Geoff Carr, Danielle Clode, Orin Courtenay, Scott Creel, Jack da Silva, Chris Dickman, Patrick Doncaster, Sean Doolan, Eli Geffen, Joshua Ginsberg, Dada Gottelli, Pall Hersteinsson, Heribert Hofer, Nick Hough, Gillian Kerby, Ian Lindsay, Malcolm Newdick, Warner Passanisi, Claudio Sillero and Rosie Woodroffe. Their work, along with that of my non-Carnivore students, reverberates through these pages, including unpublished observations which I gratefully acknowledge. At this punctuation point in my affair with Carnivores I thank my students, who have probably taught me more than I have taught them, and I fondly acknowledge those who struggled to teach me at the outset, the late Niko Tinbergen and Hans Kruuk.

It is not to diminish my thanks to others to record that above all two friends have devoted themselves to the quality of this book, and the associated films. For three years Paul Stewart and Lars Werdelin have devoted countless hours, compendious knowledge and very fine intellects to *The Velvet Claw*. I shudder to think of the errors from which they have saved me.

David W. Macdonald

INTRODUCTION

Over 65 million years, squirrel-sized creatures that scurried after insects have given rise to all modern Carnivores, ranging from bears, dogs, weasels and raccoons to cats, hyaenas, civets and mongooses. There were also descendants that did not survive to the present day, such as sabre-toothed cats, raccoons the size of bears, cat-like dogs and Cheetah-like hyaenas. This book combines reconstructed views of these predators with detailed research on living Carnivores to trace how such a diverse group of animals was fashioned from the same raw material.

Modern Carnivores range from the Tiger to the Weasel, from the Polar Bear to Fennec Fox. Their diets span the flesh-eating Leopard to the bamboo-eating Giant Panda, with every shade of variation between. These differences in diet and other aspects of their ecology have led to a diversity of lifestyles, ranging from those that lead a mainly solitary existence to those co-operating intricately within a group. This book tells two stories, that of the present and that of the past. The two are linked by one set of rules that apply now as they did millennia ago: the rules that link behaviour with ecological circumstances and in particular with feeding habits. A Carnivore's social behaviour is just as much an adaptation to its circumstances as are crushing teeth and rasping claws. The societies of Carnivores are very intricate and it is their social lives that form the main topic of this book.

The food of a modern Carnivore, the trade it plies, is mirrored in its bones, and most especially in its teeth. The scientific detectives that have worked on the Carnivore case have their specialities: the ecologists have pieced together the links between diet and lifestyle and the anatomists read a tooth like a menu. All the evidence is assembled in living Carnivores: the case for tooth-diet-lifestyle is proven. For those Carnivores that are long-dead we have only fossilised bones and teeth. But these provide clues to diet and thence behaviour. The door is open for putting not only flesh, but also lifestyles, on the bones of the past.

Exhuming the extinct ancestors of today's Carnivores is the job of palaeontologists. Detailed studies of fossil bones allow researchers to piece together the history of a lineage. Fossils can be dated, relatively speaking, by the order in which they appear in geological strata, but giving them an absolute age is very difficult. One method of ageing fossils is radio-carbon dating: because an isotope of carbon decays at a constant rate its concentration can be used like an egg-timer to age fossils. The relationships between present-day Carnivores can partly be inferred from anatomical similarities, and in particular by studying certain bones of the skull. In addition, biochemists have recently devised ingenious techniques for assessing the relationships between species, and the time since they descended from a common ancestor. One approach relies on the so-called molecular clock: changes in the amino acids making up the proteins of animal tissue occur at a constant rate. A study of these acids in one protein (for example albumen) from modern members of a Carnivore family (for example the cats) reveals differences that are proportional to the time for which each species has existed as a separate entity.

To understand the Carnivore's story one must appreciate the wonderful ideas of Darwinian evolution, and in particular the mechanism of natural selection. Each animal carries a set of genes, which make up a blueprint of that individual. The genes can be passed on to offspring through eggs and sperm. At each generation the genetic make-up of parents is reshuffled into their young, and occasionally mutations provide new characteristics. Individuals best fitted genetically to prevailing conditions have the best chance to survive, and to pass on their qualities. Thus species are not fixed entities: as circumstances change so the adaptations required to survive also change. Animals behave in ways that help to maximise the number of copies of their genes in the next generation. Most individuals do this by trying to maximise the number of their offspring that survive to breed. However, there is another possible route to the same end. Close kin share many of the same genes so an individual can perpetuate many of its genes by tending related young. This helps explain why some individuals are prepared to become non-breeding nannies, and to act in other seemingly altruistic ways. Of course, biologists observing the intricate tactics that animals adopt do not think that each individual coldly calculates the odds on its behaviour leading to more descendants, but rather natural selection ensures that individuals which behave in a manner which leads to a greater number of grandchildren will, in the end, outnumber those whose behaviour is less well adapted.

During the evolution of the Carnivores many species, and even some lineages, have come and gone as circumstances have changed over the eons. In some cases a new lineage may have ousted an old one, in others an extinction may have left a gap for a new type to fill, and old and new lineages may even pass like ships in the night. There is much in common between the forces of economics in the market place, and the forces of natural selection in evolution, as illustrated by the parable of the two potters. A father bequeathed a pot mould to his son, Potter A, and a potter's wheel to his other son, Potter B. They both made pots but A could make them faster and cheaper with his mould than B with his laborious wheel. So B, at a competitive disadvantage, was soon on the verge of bankruptcy. Then Potter B discovered that his wheel could also be used to make plates, cups, bowls and a whole range of marketable items that A could not make with his rigid mould. By diversifying, Potter B was able to stay in business alongside Potter A. Some time later tupperware pots came into vogue and the market in clay pots crashed. This was a disaster for Potter A, who could make nothing else with his mould. Potter B suffered from the loss of the pot trade but could still market his plates, cups and so on. So Potter B survived while Potter A went out of business. Then fashions changed again, and clay pots were once more in demand. Potter B noticed the gap in the market, and started making clay pots on his wheel, as he had done in the past. Similarly, lineage A of animals highly adapted to a particular niche will be able to exploit that niche more efficiently than lineage B containing less specialised members. But this does not mean that lineage A will ultimately survive nor that B will be ousted completely. If lineage B can diversify, it may flourish by exploiting other niches as well. Changes as arbitrary as the fashion in tupperware, such as shifts in climate, may adversely affect A's specialism. Specialist lineage A may go extinct, while the more generalist lineage B survives.

Carnivores have followed many different routes, none of which is superior to the others. Whether one is more awed by the Kodiak Bear's might, the Giant Panda's thumb or the Least Weasel's miniaturisation is a matter of personal choice. But awe at the Carnivore's story must be tainted by the fear that its sequel may be tragic. Life near the top of the food pyramid is precarious, and many of the most magnificent Carnivores are gravely imperilled.

CHAPTER 1

THE CARNASSIAL CONNECTION

This is a story of the Carnivores. From the mighty to the minuscule, this group of mammals all descend from a common ancestor. Today they number about 236 species, most of which have rarely been watched and never studied. Some, such as the Polar Bear, the Tiger and the Grey Wolf, are familiar. Others, like the Kinkajou, the Fossa and the Aardwolf, are as mysterious in their habits as in their names. Some swim, like the otters; some burrow, like the badgers; some climb, like the martens. Some are ubiquitous, like the Raccoon; some are imperilled, like the African Wild Dog; some are flamboyant, like the skunks. This marvellous diversity extends back over millions of years and has largely been brought about by adaptations to secure food.

Despite their name, not all these animals are purely meat-eaters. The word 'carnivore', like other nouns, can be common or proper. Animals ranging from beetles to buzzards are called carnivores because they eat other animals. Many Carnivores with a capital 'C' do eat meat, but others are omnivores and one, the Giant Panda, is a fairly strict vegetarian. But what distinguishes Carnivores from other mammals is that almost all of them have a set of scissor-like back teeth, called the carnassials, with which to shear through flesh. Some living Carnivores, such as the Giant Panda and the Aardwolf, have lost these scissor teeth, but their ancestors had them. As will become clear in this chapter, the engineering of scissor teeth was the linchpin in the evolution of the Carnivores.

Eating flesh is a luxury. Meat is easier to digest than the much more abundant and easily caught vegetation, and very nourishing. As with most luxuries, though, carnivory can be expensive, making it a high-risk, high-reward lifestyle. Catching prey is both difficult and dangerous. Once caught, converting prey flesh and bone into predator flesh and bone is simpler than the alchemy of turning greenery to flesh. Therefore, flesh specialists, such as the Wildcat and Lynx, have guts only four times their body length. Foxes and wolves, facing the problems of omnivory, have guts five times their body length. Sea Otters, whose digestion has to battle with clam shells, have guts ten times their body length. Bones are difficult to chew and digest, and the Spotted Hyaena's stomach produces copious hydrochloric acid to dissolve them. Cats can vomit voluntarily to rid themselves of piercing stomach aches from bone splinters. Herbivores have a gut pocket, known as a caecum, to digest plants, and this is lacking in flesh-eating Carnivores such as cats, weasels and mongooses. Omnivorous ones, like foxes, have a large caecum. The Eurasian Badger may be disadvantaged because it has become an omnivore, but, like all mustelids, has no caecum.

A Carnivore's diet also affects other aspects of its life, including its sociality. Some foods can be shared, others cannot; some renew within hours, others take years to replenish; some can be attained most effectively by lone animals, others by working in a pack. These differences have led to a diversity of lifestyles, ranging from animals that lead a mainly solitary existence to those whose lives depend on their companions.

Modern Carnivores are split among eight families: the civets and genets (Viverridae, Chapter 1), the cats (Felidae, Chapter 2), the dogs (Canidae, Chapter 3), the hyaenas (Hyaenidae, Chapter 4), the bears (Ursidae, Chapter 5), the raccoons and coatis (Procyonidae, Chapter 5), the weasels, otters, badgers and skunks (Mustelidae, Chapter 6), and the mongooses (Herpestidae, Chapter 7). Today these families include all the top land predators, but in the past they occupied much lowlier positions. The Carnivores have had to fight their way up from the bottom, a 65-million-year contest for the predatory crown.

To find out how the Carnivores came to supremacy we must go back 65 million years to the end of the Cretaceous period. During the Cretaceous, the dominant predators were dinosaurs, a group that also filled many other niches and had done so for 150 million years. If humans had been around at the time, an early dinosaur-watcher might have ducked for cover as a pack of brownish-striped 2-metre (6.5-feet) tall *Deinonychus* thundered past in pursuit of the graceful vegetarian dinosaur *Ornithomimus*. The plant-eater would have dodged and swerved to no avail as the 70-kilogram (150-pound) pack hunters fanned out behind it. Dodging one *Deino-nychus*, the exhausted *Ornithomimus* would have stumbled into another, which kicked out, Kung-fu style, and dragged talon-like hind claws in a disembowelling slash. *Deinonychus* would then have grasped *Ornithomimus* in long-fanged jaws and in one vault rolled on its back, pushing out with its forelimbs and kicking with its hindlimbs, like a modern Weasel, to keep the prey's flailing claws at arm's length. An instant later, the pack of *Deinonychus* would have been on the prey and, after a dismembering struggle, each would have retreated to eat its booty alone.

As in all recreations of prehistoric scenes, one can only speculate. We cannot be certain that *Deinonychus* hunted in packs, nor that it was partly camouflaged by brownish stripes and killed weasel-style. But the bones of the story are true, because the bones of *Deinonychus* and *Ornithomimus* have been preserved as fossils to show that they lived as predator and prey.

The dinosaurs are relevant to the story of the Carnivores because their mass

extinction 65 million years ago opened up a world of opportunities for creatures that had been scuttling at their feet. Among them were small, shrew-like early mammals. In dank tropical forests and swamps, the mammals and other groups diversified and adapted to fill the niches left vacant, some becoming large herbivores, others omnivores, others predators. In particular, the demise of the dinosaurs fired the starter's pistol in a competition to become top predators.

The world of 65 million years ago was very unlike that of today. The landscape looked similar to some modern tropical forests, but the position of the continents was markedly different. There had been one huge continent, Pangaea, and about 136 million years ago this had slowly begun to split up. The shapes of modern continents still show how they used to fit together, the east coast of South America snuggling against the west coast of Africa, and Scandinavia locked on to Canada, linked by Baffin Island, Greenland and Iceland. Exactly when, and in which order, the continents split up is controversial, but a popular view is that Pangaea first split into two continents, one northern, Laurasia, and the other southern, Gondwana. Shortly afterwards, just under 130 million years ago, Africa split from the southern continent, followed by India. About 60 million years ago Laurasia split along what is now the North Atlantic, leaving Greenland in the gap. North America and Eurasia drifted apart only to meet on the other side of the world where Alaska and Kamchatka have intermittently been linked by a landbridge across the Bering

Dinosaurs were not all plodding dimwits as often portrayed. Deinonychus *was a nimble predator, and may well have hunted in packs for herbivorous prey like* Ornithomimus

Straits. By the time of the dinosaurs' demise, Laurasia and Gondwana had drifted far apart and were separated by a wide ocean. For as long as the northern and southern domains remained apart, the story of predatory evolution had two separate strands.

At the time of the dinosaurs' extinction, the southern continent was home to three main groups of mammals. One of these, the edentates (the sloths, anteaters and armadillos), had no pretensions to carnivory. Their name, which means toothless, slightly overstates the case for they do have teeth, but simple and often very puny ones. The second group, the didolodonts, comprised only herbivores. It was left to the third group, the pouched mammals or marsupials, to produce the first large mammalian predators in the south. Today, their descendants span creatures from kangaroos to koalas. Marsupials produce young which are extremely premature in comparison with those of the Carnivores, ourselves and other placental mammals. Some, such as the kangaroos, stow their helpless young in pouches while others, such as the opossums, simply plug them unshakably on to their teats.

Marsupials arose in North America before the final breakup of Pangaea. Some spread over Europe and thence to Asia, but by 25 million years ago, marsupials had died out in the northern continent. Others travelled south to Gondwana between

100 and 75 million years ago, and they fared better. There they could roam lands that were to become South America, Antarctica and Australia. The ancestor of the southern marsupial predators was a small creature with a pointed snout, large ears and large eyes. It was probably similar to a small Virginia Opossum, which is a very successful rat-like omnivore in the Americas today. This ancestor was widespread by the time South America split off from the southern continent. Then Australia and Antarctica moved apart, each with its stock of marsupials on board. Drifting south, Antarctica iced over about 40 million years ago, and fossils of its fauna doubtless lurk deep under the ice. Meanwhile, Australia and South America drifted across the southern oceans as aimless arks for marsupials.

In Australia and South America pouched killers evolved in all shapes and sizes. Until a few million years ago, most of those in South America belonged to a group called the borhyaenids. One of the best fossil records was left by a medium-sized borhyaenid, *Cladosictis*, 19 million years ago. About 80 centimetres (30 inches) long, *Cladosictis* looked rather like a modern otter or marten, and lived in Patagonia. Although it was no relation to modern Carnivores, *Cladosictis* anticipated their invention of pointed canine teeth and molars fashioned as cutting scissors. It probably hunted in water as well as on land, as does the Yapok of the Andes, a member of the modern family of opossums which arose 12 million years ago. The Yapok is the only modern marsupial adapted to a semi-aquatic life. It has webbed hindfeet, and females have a rear-opening pouch which is closed by a sphincter to become a water-tight chamber during a dive. When it arose, the better-adapted Yapok probably competed with *Cladosictis* for the semi-aquatic trade. Eventually, two million years ago, these borhyaenids were rendered obsolete by modern opossums. Another lineage of pouched killers led to heavily built flat-footed animals that probably ambushed their prey. One was *Proborhyaena*, known only from part of a skull 60 centimetres (2 feet) long, the same size as that of a modern buffalo.

The pouched killers dominated South America for 30 million years during the Eocene and Oligocene, but about four million years ago many began to face competition from a new breed of predators – the phororhacoids, or thunder-birds. Relatives of modern cranes, moorhens and bustards, these flightless birds stood 3 metres (10 feet) tall astride nightmarishly massive legs, and bore mighty curved beaks that were longer than a horse's head. Doubtless the thunder-birds squabbled with vultures of the day for any carrion they came across, but since they lacked the high vantage point used by vultures to locate such food it seems unlikely that they were principally scavengers. It is easy to imagine them pounding in a terrifying sprint, wielding their massive beaks and flailing kicks that could pulverise their prey. They were the closest birds have come to wolves and, at their peak, the thunder-birds may have ousted some wolf-like borhyaenids that chased after prey. Today the thunder-birds have gone, but the crane-like Seriama that still struts the grasslands of South America is a distant offshoot of their lineage.

By about four million years ago the borhyaenids had given rise to a sabre-toothed marsupial killer known as *Thylacosmilus* or the Pouch-knife, which lived on the pampas of Argentina. It was a stocky animal, much like a modern Leopard in size and build. To out-manoeuvre large prey in open country, the Pouch-knife may have hunted in groups but it was not built for agile pursuit, having massive neck and chest muscles to power-drive its 12.4-centimetre (5-inch) sabres. How it used these weapons to kill is unknown. In 1900, one palaeontologist proposed the charming, if implausible, theory that Pouch-knives used their teeth rather like can-openers on animals called glyptodonts. These were lumbering armadillo-like mammals which invested 20 per cent of their weight in a protective bony shell and a mace-like tail. When faced with a Pouch-knife, a glyptodont would have lowered its defenceless belly to the grass, pulled its head as far into its shell as it would go and twirled its fearsome tail. A lunge from the Pouch-knife would have earned it a swipe from the glyptodont's rapidly thrashing 1-metre (3.3-foot) long mace.

Reconstructing the operation of the Pouch-knife's teeth was held up for years by misleading drawings based on the first Pouch-knife skull to be described, which had been rather distorted by fossilisation. The drawings depicted the sabre canines as diverging, an improbable design that would exert unbearable, jaw-splitting pressure on the animal as its teeth sank in to flesh. It now seems that the Pouch-knife's rapier teeth were parallel. The muscles for lowering its head were immensely strong, probably more so than those of sabre-toothed cats. Furthermore, its upper canines were rather straight, allowing a direct downward thrust to inflict a deep stab wound. So Pouch-knives may have stabbed their prey with a violent downward jerk of the head. The teeth of Pouch-knives grew continuously, unlike those of Carnivores such as sabre-toothed cats. Continuous growth made it impossible to anchor the sabres with a bulbous knob in the jaw as in the cats. Instead, a Pouch-knife's canines were deeply rooted in canals that ran through the skull to a position well behind and above the eyes. As they grew, the canines probably retained their razor edge by being sharpened on a horny pad positioned on two flanges which stuck out from the Pouch-knife's chin. Perhaps the animal drove these flanges into its prey's flanks, for use as pivots during the stabbing bite. The flanges also served as protective sheaths for the sabres, and prevented the Pouch-knives from stabbing their own chest.

While the pouched killers were consolidating their position as dominant predators in the south, a wide range of creatures competed for this role on the northern landmass. In Asia and North America crocodiles came out of the water to stalk the land about 50 million years ago. One, called *Pristichampus*, was a hoofed beast with serrated teeth, and probably preyed on early mammals. The killer crocs threatened to evolve a new dynasty of ruling reptiles to take the place of the dinosaurs. But by then the mammals had become too diverse to lapse back into the subjugation their ancestors had endured under the reign of the dinosaurs.

Pristichampus, *one of the killer-crocs which evolved over 50 million years ago as heirs apparent to the predatory dinosaurs*

The northern continent was home to mammals that differed fundamentally from marsupials. Instead of letting their young develop in a pouch, these placental mammals grew their young inside them, in a womb. They are named after a special organ in the womb, the placenta, which nourishes the young during pregnancy until they are born well developed. The first big placental mammals to contend for the carnivorous niche in the north were, surprisingly, from vegetarian stock. They came from a catch-all group of rooters and browsers called condylarths, which eventually gave rise to all modern large herbivores, from deer to elephants. Palae-ontological sleuthing has unearthed how some of these ancient vegetarians became wolves in sheep's clothing.

The first known condylarth was a small hoofed creature called *Arctocyon* that lived more than 65 million years ago. It looked rather like a modern hyrax, and probably browsed on saplings. By about 60 million years ago, some of its descendants had come to look like crosses between bears and dogs, and so are called arctocyonids ('bear-dogs') – although they are not related to either. Arctocyonids, in turn, gave rise to more dedicated flesh eaters known as mesonychids. Members of this family became dominant predators and held their position for 20 million years during the Eocene. An example is *Mesonyx* that lived in North America between 50 and 40 million years ago. It was wolf-like, with a high domed skull to power a crushing bite and notches on its cheek teeth to hold flesh torn from bone.

Arctocyon, *founder of a lineage of marsupials, the bear-dogs, that flourished 60 million years ago*

It did not have scissor-like back teeth, but the backs of its molars came close to a shearing design. Yet, like other mesonychids, *Mesonyx* still had heirlooms from its hoofed vegetarian ancestors – hooflets, albeit rather claw-like ones. One might think that a hoofed predator would be excellent at speeding after prey, but hopeless at bringing it down, lacking the crampons necessary to grab hold of it. However, for 20 million years hoofed prey were pursued and caught successfully by hoofed predators.

The condylarth 'wolf-sheep' had another heirloom from their ancestors that was to prove a fatal burden: their vegetation-grinding molars never really developed into meat-shearing scissor teeth. Their dental dexterity was limited to gripping and tearing meat off the bones of prey. Thus, most of them were eventually put out of business by two lineages of mammals that independently hit upon the innovation of carnassial scissors. Both can be traced to 60 million years ago, but it took these newcomers a long time to oust the 'wolf-sheep' completely.

One of the last 'wolf-sheep' to fall was a giant Mongolian scavenger named *Andrewsarchus*, which lived about 37 million years ago in what is now the Gobi Desert. This may have been the largest scavenging land mammal ever. Its skull was nearly 1 metre (3.3 feet) long and its body was probably over 4 metres (13 feet) long. *Andrewsarchus'* jaws were of crocodilian ferocity. It lumbered across the plains of Mongolia, scavenging from the carcasses of huge herbivores such as

The carcass of an Embolotherium *would have provided a feast for a gatheriung of* Andrewsarchus *37 million years ago.* Andrewsarchus *was one of the hoofed predators that dominated the northern continent for 20 million years during the Eocene epoch*

Embolotherium, a 4.5-metre (15-feet) long rhinoceros-like beast with a grotesque head. Lacking shearing teeth, it would have been unable to slice meat from the carcass and would have fed rather clumsily, tearing pieces off and bolting them down. Because the prey were so large, it seems likely that several *Andrewsarchus* fed side by side, perhaps operating as a loose-knit family to defend their spoils.

Before becoming extinct as predators, the condylarths set in train one of the most remarkable stories in mammalian history. About 53 million years ago a long-bodied, thinly furred animal slipped into a tropical lagoon, bobbed beneath the surface and then kicked violently as it swerved in agile pursuit of a fish. This was *Hapalodectes*, which looked much like a modern Giant Otter except that its stocky limbs were hoofed. From nose to tail it was probably about 1.3 metres (4 feet) long. Over time, its descendants became ever more adept at swimming and paid for their aquatic prowess with increasing cumbrousness on land. By 46 million years ago, this line had transformed to *Pakicetus*, a 2-metre (6.5-foot) long creature with webbed feet. It looked much like a modern seal or sea lion, with nose and eyes high on its head to keep them above water, and limbs that were well on the way to becoming flippers. About 2 million years later, the 3-metre (10-foot) long *Protocetus* had a massively

Hapalodectes *(above) was an otter-like hoofed predator whose aquatic habits were the first step to the evolution of whales*

Though toppled from their predatory supremacy, they spawned a remarkable lineage of descendants: the whales

By 46 million years ago Pakicetus *(above) was much like a modern seal. Pakicetus's legs were well on the way to becoming flippers, and its eyes were high on its head (above right)*

Natural selection has transformed modern whales unrecognisably from the hooved predators from which they descend (main picture)

Protocetus was fully aquatic, with a massively muscled propulsive tail (above)

muscled tail to provide the main propulsive force for swimming, and was probably fully aquatic. The process continued and by 35 million years ago all trace of hind-limbs had gone. A line of 'wolf-sheep' had become whales.

The animals that probably wiped out the 'wolf-sheep' on land had as their ancestor a creature called *Cimolestes*, or one very like it. More than 65 million years ago, this squirrel-sized creature scuttled in pursuit of insects at the feet of the dinosaurs, with sensitive nose whiffling and nervous eyes bulging. It looked rather like the tree shrews found today in Malaysia, and was small enough to be eaten at one sitting by most modern Carnivores. Yet all the varied creatures, from Puma to Giant Panda, that now share the name Carnivore are derived from an animal like *Cimolestes*.

Cimolestes probably supplemented its insect diet with the occasional vertebrate, perhaps killing it with a series of rapid bites as civets and genets still do today. However, its key legacy was the apparatus it used to chop the prey up. Its cheek teeth were somewhat flattened from side to side, providing the beginnings of a scissor-like action. Over millions of years these scissor teeth would be refined and honed to slice meat and sinew in what is called a carnassial shear. By about 58 million years ago, the blueprint for carnassial scissors had been bequeathed from *Cimolestes* to two separate lineages. Both modified the basic scissor-tooth design and both became successful in time. One group gave rise to modern Carnivores; but

65 million years ago squirrel-sized Cimolestes *developed the scissor-action teeth that were to lead directly to the evolution of Carnivores*

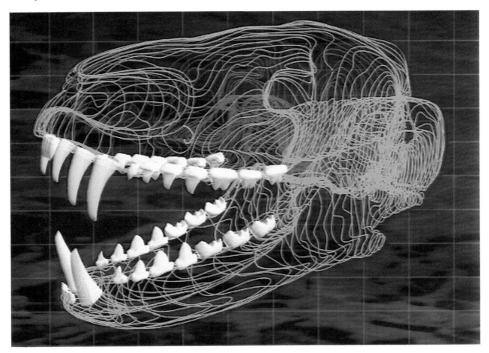

it was the other group, the creodonts, that were the dominant meat-eaters to begin with.

The creodonts flourished from about 55 to 35 million years ago. Their dominance is illustrated by fossils found in Africa, of which none were true Carnivores but 11 species were creodont. In their heyday the creodonts produced species that were dog-like, bear-like, hyaena-like and cat-like, including some with sabre teeth. There were two families of creodonts, the oxyaenids and the hyaenodontids, both of which ranged through North America and Eurasia. The hyaenodontids also colonised Africa, and contained members varying in size from that of a Stoat to a Striped Hyaena.

Among the most fearsome of the hyaenodontids was a wolf-sized animal by the name of *Hyaenodon horridus*, which lived in North America about 25 million years ago. Like modern members of the dog family, it had slim legs and walked up on its toes, indicating that it was a runner. However, its spreading toes suggest that it was built more for endurance than for speed. *H. horridus* is also noted for having a rather small brain.

Larger creodonts included the oxyaenid *Sarkastodon*, which was a rather bear-like creature, although at 3 metres (10 feet) long it was bigger than the largest true bear. This specialised bone-crunching scavenger probably fed on the huge herbivores that roamed Central Asia. The largest fossil creodont discovered, and possibly the largest mammalian land predator ever, was the hyaenodontid *Megistotherium*, which inhabited the Sahara about 20 million years ago. This creature probably weighed about 800 kilograms (1800 pounds), just a bit more than a modern male bison. A complete skull found in Gebel Zelten in Libya measured 65 centimetres (26 inches), twice the length of a Tiger's skull. *Megistotherium*'s skull may have continued growing throughout its lifetime, and it had large canine teeth and immensely strong jaw muscles to power cheek teeth designed for bone-crunching. It also had, relative to its size, the largest brain of any creodont.

By the time of *Megistotherium*, the creodonts were being pushed aside by the other group with carnassials, the Carnivores. Although the last creodont, *Dissopsalis*, plodded on in Pakistan until 8 million years ago, the Carnivores rose to become top predators on the northern continents 20 to 30 million years ago. The coincidence between the rise of the Carnivores and the fall of the creodonts strongly hints at cause and effect, but the two lineages were so similar that it is hard to see what advantages the Carnivores could have had over the creodonts. Both groups appear to have the same ancestor, and both had well developed, meat-slicing carnassials. Their fossils indicate that there was little difference in the efficiency of their teeth or their senses, although creodonts had somewhat less flexible backs. Some scientists did suggest that creodonts were outwitted by smarter Carnivores. However, while smaller than that of modern Carnivores, the brain size of creodonts relative to their body size was on a par with the Carnivores of their time.

The most convincing explanation focuses on the only known important difference between the two groups – the cheek teeth they used to scissor through flesh. The Carnivores evolved their scissors out of the most rearward upper premolar cutting against the most forward lower molar, whereas the creodonts evolved theirs further back in the mouth from the second and third upper molars and third and fourth lower molars. Consequently, the creodonts had no teeth left behind their carnassial scissors for grinding fibrous material, and this may have committed them too wholeheartedly to butchery. The Carnivores, on the other hand, had several teeth to the rear of their scissor teeth. When the opportunity arose, these back teeth could be turned to other uses, such as grinding fruit and vegetables. So while there could only be as many creodont species as it took to fill up the meat-eating niches of the world, the Carnivores could encompass specialist meat-eaters and species that flourished on mixed or even vegetarian diets. Consequently, Carnivores could become more abundant and diverse than creodonts. In fact there were 45 genera of creodonts, and there are now 98 living genera of Carnivores. If this explanation has any validity, then, ironically, it was their ability to diversify from carnivory that secured the Carnivores their success.

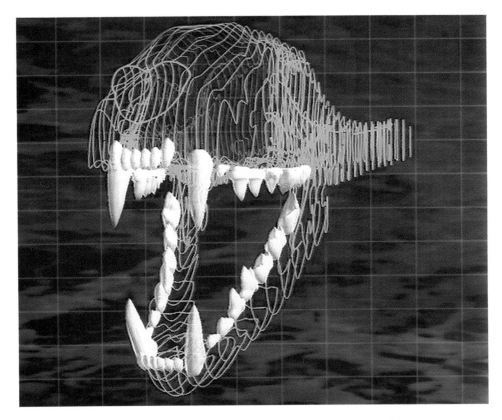

Hyaenodon *(above and opposite), a creodont, used its rear-most teeth as carnassial scissors*

There must also have been a change in circumstances that affected the specialist creodonts more than the Carnivores, and this may have been a dramatic shift in climate that occurred just over 30 million years ago. The evidence comes from the shape of fossil leaves: leaves from warm, rainy climates tend to have smooth margins and pointed 'drip-tips'; those from colder places are often smaller with serrated margins. The proportions of these types in fossil forests changed radically at the end of the Eocene. Previously, Laurasia had been warm, with year-round rain and little seasonal variation in temperature. Evergreen tropical rainforest extended 20–30 degrees further north than today, and even at 60 degrees north, in Alaska, mean annual temperatures reached 22°C (72°F). Afterwards, mean temperatures were 12°C (22°F) lower, and temperate deciduous broad-leaved forest predominated. The shift was probably caused by a change in the inclination of the Earth's axis and thus the amount of sunshine falling at given latitudes. Its effects varied at different latitudes, but generally the diversity of plants fell as many failed to make the change. Perhaps the initial loss of plant diversity also reduced the prey for creodonts, whereas the greater seasonality produced regular gluts of fruit and insects which could be exploited by the versatile Carnivores. Cycles in climate continued at about 9.5-million-year intervals, with temperatures peaking at 7–9°C (13–16°F) above the minima about 25, 15 and 5 million years ago. As will become clear in later chapters, at least the first and last of these peaks coincided with major changes in the history of the Carnivores.

The early Carnivores, known as miacids, were unspectacular. Between 60 and 55 million years ago they were mainly lithe, tree-dwelling hunters in the lush forests of the time. Two types of miacids lie at the base of the Carnivores' story: the vulpavines and the viverravines. The vulpavines lived largely in the New World and may have resembled modern Pine Martens, athletic hunters of the canopy. The viverravines were based in the Old World and they probably looked much like modern genets, which are predators of small vertebrates. Genets, along with civets, are members of a modern Carnivore family, the Viverridae. Although the viverrids are probably no more ancient than other living Carnivore families, they are the most similar to primitive Carnivores in appearance. This is because many have never left the habitats occupied by early Carnivores, and so are adapted to similar lifestyles. The viverrids are thus a window on the past but, as so little is known of modern viverrids, the glass in that window is somewhat opaque.

Civets and genets are found only in the Old World. The oldest known are *Semigenetta* from Europe 24 million years ago and its African contemporary, *Kichechia*. The 35 modern species of viverrids are generally tree-dwellers, with five toes on each paw, and cat-like retractile claws. Many are spotted, and they usually have long tails. Modern viverrids vary in size from the 20-kilogram (44-pound) Fossa to the 600-gram (1.3-pound) Spotted Linsang, a sinuous nocturnal tree-climber inhabiting Borneo.

The viverrids have two particularly notable features. First, the skin of the outside rim of each ear is folded to form a little purse, the function of which is unknown. Second, most viverrids have an elaborate scent pouch, called the perineal gland, which lies between the genitals and anus. Long ago, their glands brought viverrids to the public eye, or at least to the public nose, and people began scooping out from the glands the thick, greasy, yellowish secretions known as musk or civet. Shakespeare wrote in *As You Like It* that 'Civet is of baser birth than tar, the very uncleanly flux of a cat'. Its scent at high concentrations can be nauseating, but it is attractive in minute traces. More importantly for humans, it is an exaltant, enhancing and prolonging the fragrances of aromatic perfumes.

Civet secretion is reputed to reduce perspiration, cure skin disorders and, like everything else, act as an aphrodisiac. An African Civet accumulates 4–19 grams (0.14–0.67 ounces) of musk a week in its deep muscular pouch. In Ethiopia, a civet farmer may keep as many as 60 wild-caught male African Civets. Trade in their secretions is very ancient; King Solomon imported civet musk from Africa in the tenth century BC. Although synthetic alternatives now exist, the trade still flourishes in several East African and Oriental countries. African Civets have been introduced to some impoverished areas of Ethiopia to stimulate the economy.

Miacids were the first Carnivores, and used the fourth upper premolars and first lower molars as carnassial scissor teeth, thereby leaving their rear-most molar teeth free for other roles

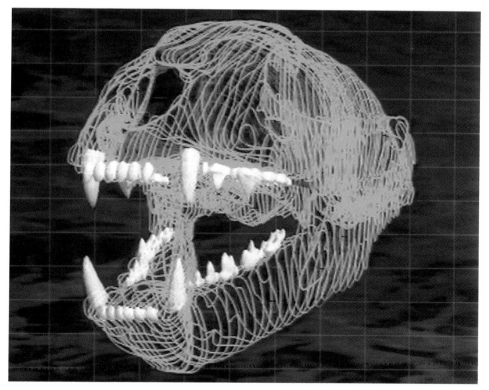

Viverrids are divided into four subfamilies which differ in terms of scent pouches. The three species that comprise the Madagascan subfamily have no scent pouch at all, although one of them, the Fossa, may be seen embracing tree trunks while gyrating its genitals in a way that presumably spreads scent. Perhaps such behaviour led to the evolution of scent pouches. The second subfamily comprises the banded palm civets and otter civets, and these animals have rather small pouches. In the third subfamily of palm civets, the gland lies within a simple long fold of skin and produces only a thin film of civet. It is present in both sexes of all species of palm civets except the male Small-toothed Palm Civet. The true civets (which include the genets) have the most ornately cavernous pouches, with the exception of the linsangs which, puzzlingly, have no pouch at all. True civets scent-mark and defecate at dung heaps or 'civetries' at territorial boundaries and along trails.

From the fossil evidence, the teeth of the early Carnivores were very similar to those of the African Palm Civet. This means that their food and social life were probably also much like the African Palm Civet's. Despite its name, this African animal is probably not a member of the palm civet subfamily, all of which are confined to Asia. It has much sharper scissor teeth than true palm civets and, although its affinities are still unknown, it may be closer to the true civet subfamily. Like most true civets, African Palm Civets are primarily nocturnal foragers, gleaning prey as they meander through the canopy, occasionally taking fruit in addition to a staple diet of small vertebrates, eggs and insects. Except for fruiting trees, these food items are too sparse to be shared, and once harvested, take a while to replenish. These disadvantages to sharing a territory probably explain why African Palm Civets are solitary creatures. An adult male occupies a territory of over 100 hectares (250 acres), and regularly scent-marks trees along the border, and fruit-bearing trees within. Up to three females live within a male's territory, each with her own plot of ground shared only with the male and her sub-adult daughters. Other males, perhaps adolescents, squeeze into the territory but maintain a low profile. Male and female African Palm Civets socialise only rarely, seldom foraging in the same tree, doubtless to avoid getting in each other's way. However, they keep track of one another through eerie siren-like calls, sometimes performed as a true duet.

Most other viverrids are also primarily nocturnal foragers, but only snippets are known of their tastes. The Feline Genet, for example, raids birds' nests. Forest Genets lurk in cave mouths to swipe passing bats as they leave their roosts, and then meander off to feed on the sweet nectar of bat-pollinated trees. Common Palm Civets are partial to a drop of toddy, the juice tapped by people over much of southern Asia to ferment as palm wine. Their drunken habits have earned these civets the popular name Toddy Cat. Amongst the 34 other foods Toddy Cats have been recorded eating in Java is the fleshy husk of coffee beans. The civets expel the

The African Civet's scent glands produce a valuable ingredient of expensive perfumes

The Genet (above) lives today much like some of the earliest cat-branch Carnivores

The Small-toothed Palm Civet (opposite) forages in trees for small vertebrates, insects and fruit in Southeast Asia

undigested kernels in their faeces and, when collected and ground, these passaged beans produce, so it is said, the finest coffee possible. Several civets, such as the Celebes Civet and Small-toothed Palm Civet, eat a lot of fruit.

Some of these foods might be more easily shared than the prey of strictly carnivorous viverrids. For example, a civet living largely off tropical fruit could hold a territory in which there are always at least some trees in fruit, each providing enough food for several adults. In such circumstances a male might better maximise his offspring by holding a territory with a mate and helping to tend their young. Unfortunately, insufficient is known of viverrids' social lives to decide whether fruit-eaters are more gregarious than those dedicated to less shareable vertebrate prey.

The males and females of most viverrids are about the same size, but where there is a difference it is the female that is larger. One would have expected the reverse, at least in those species such as the African Palm Civet where one male fights to

monopolise several females. Perhaps the sexes are usually similar in size because their need to tiptoe along thin branches in pursuit of food made it impossible for males to become bigger, and so fiercer. The most extreme size difference occurs in the Binturong, a type of palm civet in which females are reputedly 20 per cent heavier than males and have rather masculine genitals. Binturongs have departed most from the usual viverrid mould. They have a shaggy black coat, each hair tipped yellow or white. As a Binturong climbs, its scent pouch drags on the branches leaving a perfumed smear said to be reminiscent of cooked popcorn. In the trees, a Binturong's long prehensile tail acts as a fifth limb, allowing it to walk upside down, using its forelimbs to pull fruit to its mouth. Only one other Carnivore has a prehensile tail: the Kinkajou (see page 155).

Viverrids are often said to illustrate the ancestral Carnivore design of arboreal, nocturnal hunters, that has been adapted by the other modern families. Yet viverrids themselves have contenders for several Carnivore trades. For example, the African Civet is a robust, 12-kilogram (26-pound), dog-like creature with an intimidating mane patterned in dots and dashes. It is the only viverrid that, like the dogs, has non-retractile claws. The Otter Civet of Southeast Asia and western Indonesia lies in ambush for prey in water. It is adapted for an aquatic life with dense fur, long whiskers and nostrils that open on to the top of the snout and can be closed, as can the ears. It also has dual-purpose dentition. Its premolars are unusually large, and sharp and blade-like to cope with slippery fish and frogs. Its molars seem to be from a quite different animal, being flattened to pulverise molluscs and other shellfish.

The three species in the Madagascan subfamily, the Fossa, Falanouc and Fanaloka, are particularly diverse. This is because the only other Carnivores sharing their island are some rather weird mongooses, and the viverrids have adapted to roles that other Carnivores might have filled had they been there. Apart from all lacking a scent pocket and being very rare, the Madagascan trio have little in common. A 20-kilogram (44-pound), flat-footed tree-climber with partially webbed toes, the Fossa is so like a cat that it was previously misclassified as one. Its diet includes 10-kilogram (22-pound) lemurs, which are probably the biggest prey taken by any viverrid. Perhaps the most extraordinary thing about the Fossa is that females show genital mimicry of males – that is, they appear to have penises. Fossas indulge in marathon copulations, sometimes lasting 165 minutes, with the male gripping the female's neck throughout. The second Madagascan viverrid, the 4-kilogram (9-pound) Falanouc, most closely resembles a badger, having an elongated snout and body, and non-retractile claws. It feeds mainly on invertebrates. The 2.2-kilogram (5-pound) Fanaloka looks more like a fox, has retractile claws and feeds

The arboreal Binturong of Borneo is unusual amongst Carnivores in having a prehensile tail, and because females are bigger than males

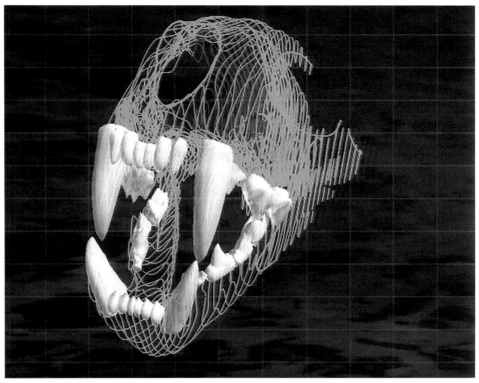

The Madagascan Fossa's skull (above) reveals teeth that are remarkably like those typical of the cat family

The Fossa (opposite) is actually a viverrid, but in the absence of cats it has adapted to their trade in Madagascar

primarily on small mammals, reptiles and amphibians.

Departing from the usual viverrid tree-dwelling role has allowed the Madagascan trio to exploit different niches. For similar reasons, the early miacids evolved different forms. About 55 million years ago these early arboreal Carnivores gave rise to two main branches: the cat-branch and the dog-branch or, more formally, the Feloidea and the Canoidea. The dog-branch grew from the vulpavines of the New World, whereas the cat-branch grew from the viverravines of the Old World. Each branch has since diversified to produce four of the eight modern families that are the topic of this book. The cat-branch sprouted the four 'feliform' families: civets (Viverridae), cats (Felidae), hyaenas (Hyaenidae) and mongooses (Herpestidae). The dog-branch gave rise to the four 'caniform' familes: the dogs (Canidae), bears (Ursidae), raccoons (Procyonidae) and weasels (Mustelidae). A ninth Carnivore family, the half-dogs (Amphicyonidae) also grew from the dog-branch. This died out about six million years ago, probably due to competition with true dogs.

Before the appearance of the modern families, the dog-branch and cat-branch evolved separately in the New and Old Worlds, respectively. However, by the early Oligocene, 30 million years ago, this tidy partitioning was shattered. The Bering

landbridge opened up between America and Eurasia, and representatives of each branch crossed. Although the true dogs remained in North America until about six million years ago, other dog-branch families became more common in Eurasia than in America. Most cat-branch Carnivores remained in the Old World, but the true cats subsequently crossed to America. The two branches of the Carnivores soon came face to face and, as will become clear in the following chapters, their meetings were often as explosive as encounters between modern cats and dogs!

Evolution had also been continuing in parallel in the northern and southern hemispheres while their continents drifted as islands apart. By the late Miocene, seven million years ago, all the modern families of Carnivore had arisen in North America and Eurasia. The basic types of Carnivore we know today all existed, although no modern species were on the scene. In the south, the marsupial mammals were still the dominant predators.

During the long separation of north and south, a few species of mammal had crossed the divide, flying, floating on rafts or, perhaps, simply swimming with favourable currents in the seas that divided the continents. Few survived the journey, and even fewer bred once there, but in the millions of years that the continents remained separated, a tiny fraction were lucky. One of these was a North American Carnivore, *Cyonasua*, which made the crossing about 7 million years ago. It was related to modern raccoons and coatis, and its name means dog-like coati. The first individual may have been a pregnant female which scampered aboard a floating raft of vegetation during a great flood in what is now Texas, only to find herself washed up days later on a foreign shore. She would have found a world quite different from the one she had left. Giant flightless thunder-birds ran down prey on plains where huge camels grazed. Then, as now, omnivorous Carnivores like *Cyonasua* were adaptable. She would have borne her young, and learnt to feed them on prey she had never caught before and protect them from predators her forebears had never faced.

These creatures and their descendants flourished for 5 million years in South America. The idea that a handful of wayfarers could found a whole dynasty is awesome, yet there is evidence of this occurring even in the history of humans. From the genetic similarities between human populations, scientists estimate that fewer than 100 of our ancestors colonised Eurasia from Africa and, subsequently, only 10 crossed to North America to found the native peoples there.

One of *Cyonasua*'s descendants was the size of a Giant Panda, the gentle giant, *Chapmalania*. Its sheer size and strength were probably enough to protect *Chap-malania* from the pouched and beaked killers, whilst it grazed on fruits and shoots. Sadly, neither *Chapalmalania* nor any of the pioneering *Cyonasua*'s other descendants survive today. They were wiped out in an event that changed their world irrevocably and devastated their southern contemporaries, the Pouch-knives.

The marsupial Pouch-knives of the south had almost perfect doubles in the northern hemisphere: the sabre-toothed cats such as *Smilodon* (whose name means 'knife-tooth'). The two groups were completely unrelated and their uncanny resemblance illustrates the pious idiom that there is only one way to do a thing, and that is the right way. The Pouch-knife and *Smilodon* filled similar niches on their respective continents and developed similar adaptations to succeed at their trade. These two sabre-toothed predators stood on either side of the divide, like mirror images, as the two Americas drifted closer.

Two million years ago, volcanoes welled up from the ocean bed and pushed their way to the surface in a line from North to South America. First as stepping stones, then as a bridge, they formed a Central American land-crossing between the two continents. For the first time since the break-up of Pangaea, there was an easy way for animals of the northern and southern continents to meet. At first the creatures of north and south seemed evenly matched. A couple of dozen genera of yankee animals invaded the south, and a dozen southern genera penetrated north. This ratio was roughly proportional to the sizes of the two faunas and so represented a fair swap. However, it was the yankee invaders that thrived. The North American incomers have generated many more species in South America than have the original residents, and the latter have suffered more extinctions. As a result, half of all mammal genera in South America today are descendants of the early yankee invaders. In stark contrast, only armadillos, opossums and porcupines from the original southern immigrants remain in North America, and even they have not diversified much.

The great interchange of species between the Americas clearly turned into a rout. The sabre-toothed cats obliterated the Pouch-knives. Before the Central American isthmus rose up, not a single member of cat, dog, bear or weasel families had set foot on South America. Soon that continent was supporting more species of dogs and cats than any other. Many types of opossum fell in the face of these incomers. Although a true Carnivore, the early settler *Chapmalania* was ousted by bears that crossed the landbridge. A thunder-bird, *Titanis*, did invade North America and lived for a while in Florida before becoming extinct, but the rest of the thunder-birds down south were wiped out.

More recently, history has repeated itself. Meddlesome humans introduced Carnivores to Australia, producing a similar effect on the local marsupials to that of the great interchange in South America. Nonetheless, some pouched killers still hang on there. The Tasmanian Devil was probably ousted from mainland Australia by the Dingo, transported there by people 8000 years ago. The Devil still survives in Tasmania, an island never reached by the invading dogs. This sturdy 5–9-kilogram (11–20-pound) scavenger also hunts small mammals. In the past, it probably fed on the leavings of the much larger Tasmanian Wolf or Thylacine, which became extinct on the mainland about 3000 years ago, again through competition with the Dingo. The Thylacine was decimated in Tasmania in the early part of this century by bounty hunting and perhaps by disease, and has not been definitely sighted since the 1930s. The quolls are probably the fiercest of the surviving predatory marsupials. Their likeness to polecats led them to be called 'native cats'. The Eastern Quoll's range has decreased in recent years: this is attributed to competition with Red Foxes introduced in 1871 for sport and in a vain attempt to control the rabbits Europeans had introduced earlier.

Australia also had a candidate at one time for the King of Beasts, the Marsupial Lion *Thylacoleo*, which bowed out about 10,000 years ago. The 1.25-metre (4-foot) long *Thylacoleo* stemmed from a line of dedicated vegetarians and its nearest surviving relatives are the phalangers, endearingly cuddly, forest-dwelling fruit-eaters. However, there was nothing at all cuddly about the Marsupial Lion. Unlike the Pouch-knives. *Thylacoleo* had made the carnassial connection: its scissor teeth were as finely engineered as those of a true Carnivore. In addition, these carnassials were embellished with serrated borders. Its canine teeth were adapted to exert a choking grip, and it also had huge claws on the inner edge of its forepaws. This adept tree-climber preyed on ground-dwelling kangaroos. Probably, like the modern Leopard, it dragged its prey into trees and out of reach of its scavenging contemporary, the Tasmanian Devil.

Scientists have tried to explain why pouched killers put up such a poor show against the Carnivores. Their limb proportions indicate that the extinct pouched killers were slower than their Carnivore counterparts, but it is hard to believe that they could not have evolved longer limbs if that was what was necessary to compete successfully. One possibility is that the early denizens of South America were less beset by evolutionary catastrophes, such as climatic changes, than were their contemporaries in North America. Consequently the yankee victors were of more vigorous stock and had been, as palaeontonologist Stephen Gould put it, 'tempered in a hotter evolutionary furnace'. However there is no evidence to support this otherwise appealing hypothesis.

As with the demise of the creodonts, the most promising proposal relates to teeth. The pouched killers, like modern marsupials, had no milk teeth. Milk teeth allow placental mammals to swap their infant teeth for a new set suited to the needs of adulthood. Young Carnivores have milk carnassial teeth, which are jettisoned when the permanent scissor teeth erupt. The molars to the rear are then used for other purposes (as in the dog family) or reduced (as in the cat family). In contrast, the ancient pouched killers opted for a sort of molar conveyor belt: molar teeth erupted in turn, each taking on the scissor role before being pushed forwards in the jaw by the eruption of the next molar in the sequence. Thus, all the molars in the mouth of a pouched killer started as scissors, and none was free to develop other functions.

The teeth of the Common Palm Civet or Toddy Cat are very like those of the first Carnivores. Its staple diet is small vertebrates, insects and eggs, supplemented by fruit

So, like the creodonts, the pouched killers probably had to specialise in carnivory. The Carnivores, though, could adjust their diet when circumstances changed, a flexibility that may stem partly from their adaptable molar teeth.

And so the Carnivores became the dominant mammalian predators, and have remained so for a few million years. However, this does not mean that they are superior to all the other predators that flourished in the past. For example, the fact that the dinosaurs became extinct might imply that they were evolutionary failures. On the contrary, while they flourished they comprised a diversity of large predators rivalling that of the modern Carnivores. Furthermore, they remained the dominant predators for nearly 150 million years, so we will have to wait another 100 million years before concluding that the Carnivores have done better.

Among more recent contenders, the creodont *Hyaenodon*, of 20 million years ago, could probably do everything a Lion does today, and some assert that the extinct Dire Wolf of 5 million years ago was more intelligent than any living Carnivore.

The problem facing predators, and all creatures, is that the rules of the game change with time. Being the fastest, smartest or strongest only helps during phases when the rules favour these qualities. Today's Cheetah is the fastest land mammal, yet fares very badly in competition for prey with the markedly slower Spotted Hyaena and Lion. Indeed, a slight shift in the African climate could favour longer grasses that would put the Cheetah out of business. If the new were always better than the old, then a modern Lion transported back in time 20 million years to the era of the first big cats might be expected to do rather well. In fact, the prey at that time were mostly giant herbivores, which would have dismissed the modern Lion with a kick. But they had good reason to fear the slow yet powerful sabre-toothed cats whose teeth were custom-built for slicing thick hides. While the Lion's conical canine teeth are well-suited to tackling the grazers of today, it is likely that sabre teeth will once more be in vogue if thick-skinned herbivores ever dominate the world's grasslands again. Unfortunately conditions do not change predictably, so we cannot foretell the shape and form of the predators that are to come.

Family Tree: The Carnivores

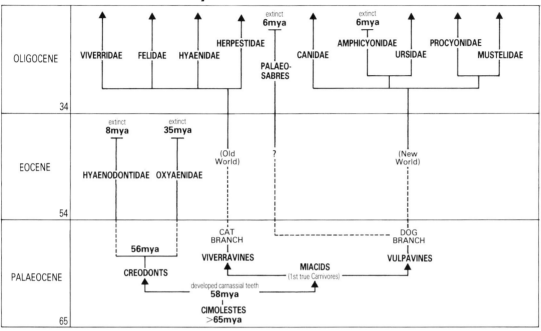

All modern members of the Order Carnivora are linked by common descent from the Miacids, the first true Carnivores. The ancestry of modern Carnivores has been traced through the fossil record to its roots almost 60 million years ago. Today, modern biochemical techniques have further unravelled the relationships between families of Carnivore and, within each family, between species (see page 8). There are still many gaps in the story, but the likely links between the eight living families are schematised in the family tree above. Similar family trees are given for the cats (page 77), dogs (page 117), hyaenas (page 145), bears and procyonids (page 179) and mustelids (page 217) – the evidence for the viverrid and mongoose families is so fragmentary that no tree is presented for them. These trees indicate common ancestry by a series of branching arrows, the likely date of separation from a common ancestor of lineages leading to modern species being indicated on the scale of geological epochs (given in terms of millions of years ago, mya, in the left margin). For example, on page 145, the line leading to the Aardwolf splits off in the Miocene: this does not mean that Aardwolves have been unchanged since then, but rather that the lineage leading to modern Aardwolves have been distinct for 10 million years. As a simplification for diagrammatic purposes, modern species are generally named in the Holocene, with no detail on their ages. Where links are particularly speculative the lines are dotted, where lineages are extant they terminate in an arrow-head, where they are extinct they terminate in a T-bar. The family trees are incomplete, but indicate the position of key players and events in the text. Their purpose is to provide a quick reference to the strands of the story (see page 42).

Strands of the Story

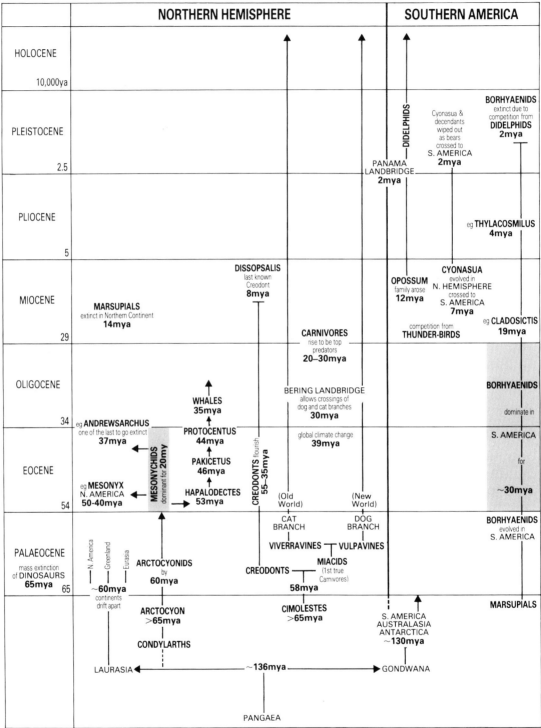

CHAPTER 2

Sharpening the Tooth

Amongst the Carnivores there is one family, and only one, whose members are all specialised killers: the cats. For them omnivory is not an option. From the largest, 360-kilogram (750-pound) male Siberian Tiger to the smallest 1-kilogram (2.2-pound) Black-footed Cat, they have traded the long faces of their ancestors for high domed skulls and short snouts which provide anchorage for muscles that power a lethal bite. The scissor-like carnassial has played a crucial part in the evolution of all Carnivores, but it is the cats that have sharpened it to the finest blade.

The cats' trade relies on stealth and ambush. Its first known practitioners evolved in forests from ancestors that were probably much like modern genets. These hunters of the canopy are not true cats but members of another cat-branch family, the viverrids. As the omnivorous ancestors of cats descended to the ground, some took to lying in wait for prey in forest glades. Doubtless, like another modern viverrid, the Fossa, these early cats were at home in the branches and on the ground. The Madagascan Fossa shows how the basic Carnivore body plan of 40 million years ago, which lives on in genets, can be converted to that of a cat. Cast in the mould of a cat, the Fossa uses dagger-like canines and retractile claws in the hunt for birds and lemurs (see page 33).

Exactly when the story of the cats began depends on what qualifies as a cat, which is controversial. The problem is exacerbated by the fact that early cats lived in forests, where fossils rarely form, and there are few remains of the middle ear bones that distinguish Carnivore families from each other. The earliest cat-like Carnivores belong to two lineages that lived about 40 million years ago in the late Eocene jungles of North America. One lineage was represented by *Hoplophoneus*, and the other by *Dinictis* which inhabited South Dakota and was rather like a modern Serval. These lineages shared a trait that is now a hallmark of modern cats – stabbing canine teeth – and most of them, like *Hoplophoneus*, were sabre-toothed. So convincingly cat-like are these fossils that they were at first called the palaeofelids and nominated as the first cats. However, their residence in the New World proved a stumbling block. The cat-branch of the Carnivores is otherwise solidly rooted in the Old World and it was the dog-branch that first developed in the New World. So the first practitioners of the cat trade may actually have been dog-branch Carnivores. Consequently it is less confusing to rename them the palaeo-sabres. Forest clearings had opened up, providing a niche for ambush predators, and the palaeo-sabres were adapted in cat-like ways to fill this niche.

Another possible candidate for the first true cat is *Proailurus*, which lived in Europe 30 million years ago. *Proailurus* was reasonably cat-like and a member of the cat branch, but it may have been just a rather catty viverrid, like the Fossa. Since two other cat-branch families, the hyaenas and mongooses, both originated in the early Miocene, about 22 million years ago, perhaps cats also arose then. Certainly, by 20 million years ago *Pseudaelurus* had evolved and its bones identify it indisputably as a cat. Between 20 and 16 million years ago *Pseudaelurus*'s kind crossed to North America.

Pseudaelurus and other early cats were medium-sized ambushers of small vertebrates. That is still the lifestyle of the oldest surviving cat lineage, the ocelots, which split off 12 million years ago. Today this lineage is represented by the Ocelot, Margay, Geoffroy's Cat, Tiger Cat, Pampas Cat and Kodkod in South America. Their ancestors probably arose in south central Asia and travelled east across the Bering landbridge between five and six million years ago, before crossing to South America about two million years ago. All six modern species are thought to have arisen after the lineage had reached South America. Some members of this ancient lineage have clung to their ancestral forest habitat. The Margay, for example, is so completely adapted to tree-dwelling that it can swivel its ankles through 180 degrees in order to hang on while descending tree trunks head first. It can also jump vertically 2.5 metres (8 feet) from a standing start. Other members have adapted to different habitats, such as the Pampas Cat which, as its name suggests, has left the forests to hunt the great grasslands of South America. The Andean Mountain Cat is an enigma: the bones of its ears are different to those of all other cats; whether it belongs to the ocelot line, or to another branch of the family is unknown.

In addition to the six South American species, there is a possibility that a seventh ocelot survives in the Old World. The Iriomote Cat was discovered in 1965, living on the forested island of Iriomote, east of Taiwan. It is the rarest cat in the world, with an estimated population of 80. Although it looks much like other Eurasian small cats, there has been speculation that it is the last representative of the ocelot lineage in the Old World.

About two million years after the ocelots branched off, the small wildcats arose, one of which was eventually to become our domestic pet. The first of this wildcat lineage was the Pallas' Cat of the steppes of Asia, soon followed by the Jungle Cat, European and African Wildcats, and two tiny, desert species, the furry-footed Sand Cat of the Middle East and the Black-footed Cat of southern Africa. The Chinese Desert Cat (which probably lives on mountains, not deserts) may also belong to the

The Margay of South American forests is a representative of the Ocelot-lineage in the cat family

wildcat line. Among the clues to the early separation of Pallas' Cat is the intriguing fact that it alone of the wildcats lacks an ancient germ known as a retrovirus. This germ must have infected the common ancestor of the other species more than six million years ago, but after the Pallas' Cat split off.

Irrespective of their ancestry, most small and medium-sized cats lead social lives that have probably remained much the same since the days of *Pseudaelurus*. The basic pattern is illustrated by the European Wildcat: each female occupies a 200-hectare (500-acre) territory, three or more of which are spanned by the territories of individual males. Bobcats in Idaho have a similar arrangement, a male's territory being two to five times larger than each female's. This might be called exclusive territorial polygyny because, rather than roaming promiscuously, males try to monopolise the females within their territories.

This system is an adaptation to their diet. As with carnivorous civets (see page 28), the prey of these cats are too small to share, and require no cooperation to capture or defend. Cats sharing hunting grounds would be disadvantaged because one arriving at an area another had hunted recently would find prey hiding and extra-alert. Furthermore, small vertebrate prey take a while to breed again if several are killed, so a cat whose patch was hunted out by rivals could face a long lean period. The result is that many female cats live alone in relatively large territories.

However, Jungle Cats are reputed occasionally to live in pairs, and Fishing Cats have been photographed in pairs.

Male cats, large or small, rarely if ever bloody their paws for the sake of their kittens. The burden of feeding the kittens is entirely the female's. However, Lions will withdraw to allow their cubs to feed at a kill and male Tigers may feed alongside their cubs. It seems that male cats have opted out of kitten care and instead maximise the numbers of their offspring by fighting for a large territory in order to monopolise as many females as possible. Since prowess in battle hinges largely on size and strength, there has been strong evolutionary selection for big males. Amongst small species they may be 20 per cent heavier than females, and 50 per cent heavier in big species.

Cats have acute hearing to help them locate their prey. A Serval, its ears swivelling like radar dishes, can hear the incautious nibbling of rodent teeth. To judge by the amount of their brain devoted to processing scent, odours are less important to cats than to most other Carnivore familes. The cat with the least developed sense of smell is the Fishing Cat, presumably because this sense is useless for fishing. Most

The Fishing Cat occurs in forests and swamps from India to Indonesia, and by feline standards has a relatively poor sense of smell

of all, though, cats rely on their sight. Some, like the Cheetah, have an enlarged area in the brain presumed to process complex images. The nocturnal eye-shine of a cat's eyes is due to a special light-reflecting membrane, the tapetum lucidum, which allows them to hunt by starlight.

For ambush and stalk, cats need to be camouflaged to the black-and-white vision of their prey. Lives depend on detecting striped Tigers in tall grass, the indistinct rings of Snow Leopards against rock-strewn cliffs, Ocelots in dappled forest shade, dark Jaguarundis in dank undergrowth, and Lions stalking in dry savanna. In some cats, different colour forms have developed to suit different habitats. While the proverbial Leopard cannot change its spots, real Leopards can obscure them: the black panther, originally thought to be a distinct species, is in fact a black-coated Leopard. In certain lights, the spots shimmer through its black fur. This melanism, caused by a recessive gene, is said to be most common in Leopard populations in forests, in mountains and in Asia. In the Malay peninsula where Leopards live in dense forest, as many as 50 per cent are black. Similarly, black Jaguars and Servals are quite common, and black Geoffroy's Cats and Bobcats crop up occasionally. Another colour variety once thought to be a separate species is the spectacular striped 'King Cheetah'. Breeding experiments have revealed that it is a rare mutant of the normal Cheetah, rather like that which produced the 'striped tabby' form of the Domestic Cat.

With paws committed to killing and climbing, not digging or running, cats have a short attack range. The desert Sand Cat swipes at a gerbil, the European Wildcat pounces on a rabbit and the Ocelot leaps to pluck a bird from the air. Ambushers are unselective, killing the unwary but not necessarily the infirm. To get close enough to attack, cats stalk their prey. The feline stalk – head and body held low, intermittent fast approaches when the prey is not looking, the final sway and explosive leap – is almost universal. *Pseudaelurus* probably stalked this way 20 million years ago. Once grappling with prey, cats kill efficiently, as any one-to-one assassin must. The faster the prey is disabled, the less likely the attacker is to sustain a crippling injury. The canines of small cats function as long, sharp daggers, stabbed into the prey's neck. They are exactly the right width to avoid the vertebrae and wedge in the gaps between them, prising apart the bones to sever the spinal cord. The canine daggers can feel their way to the killing spot, being well supplied with nerves. Combined with very rapid testing bites, this enables the cat to chart the correct path at lightning speed. The stabbing power to wield these dagger teeth necessitates bulging jaw muscles working on a short, compact muzzle. As a result, modern cats have no space left for several cheek teeth that once fitted into the longer muzzles of their ancestors. So while most modern civets and mongooses still have 44 teeth, the cat-like Fossa has 32 and most cats have only 30 teeth. Pallas' Cat and

(Overleaf) A North American Bobcat's 'iron claws' bared from their 'velvet scabbard'

the Lynx have shrunk the number even further, having 28 teeth. Cat jaws have a large space behind each canine, due to the loss of one upper and two lower premolar teeth. This space, combined with the curvature of the upper jaw, ensures that the canines stab as deeply as possible into the prey. The cutting edge of the scissor teeth makes them the ultimate slicing tools, and their tongues are coated with sharp spikelets which rasp flesh from bone.

In order to wedge between the vertebrae of their prey, the canines of cats feeding on larger prey need to be more widely spaced, broader and stronger than those of cats feeding on smaller prey. Indeed, canine teeth fit cat to prey as hand to glove; and this may help to minimise competition between species and between the sexes of one species. In Israel, Caracal, Jungle Cat and Wildcat all occur together and in each species males are bigger than females. There is overlap between the species and sexes in the sizes of their skulls and carnassial scissors. However, the diameters of their upper canines differ in a clear-cut sequence, allowing them to be ordered in neat, equal-sized steps from the smallest, female Wildcat, to the largest, male Caracal. A similar pattern applies where these three species meet in India's Sind Desert, but there each canine is smaller. The dagger teeth are apparently pushed down the size scale by the addition of a fourth, larger competing species, the Fishing Cat. The same regular spacing between species, and between sexes of one species, applies to the size of the gape of Jaguar, Puma, Ocelot and Jaguarundi. The Jaguarundi and Margay are an exception to prove the rule: their gapes are similar but they use different habitats, the Margay being much more arboreal than the Jaguarundi. Where similar-sized cats do compete, the relationships between them are strained: Lions and Tigers will kill Leopards given the chance, and Lions will seek out and kill Cheetah cubs.

The pet kitten stalking a ball of wool displays traits shared by cats from Kodkod to Clouded Leopard. Batting with forepaws in play develops into swatting a rodent with a swipe whose origins lie in movements used for tree-climbing 20 million years ago. The first cats to come down to earth capitalised on the heirlooms of their forebears. They used the flexible forelimbs and retractile grasping claws of tree-climbing ancestors to pinion their small prey ready for the dagger-thrust to the neck. In time, some cats' bodies were adapted to climb large prey as their ancestors had climbed trees. These larger, stronger cats put the climbing irons of their ancestors to predatory use and became powerful grapplers, wrestling larger prey into position for a killing bite.

However, delivering a *coup de grâce* to large prey, with muscular necks and massive vertebrae, requires a special weapon. Large herbivores had been around since 50 million years ago, and so the cats were neither the first to face this problem, nor the first to solve it. The solution, sabre teeth, was invented independently by at least four mammal groups, on three continents.

A creodont, *Machaeroides*, was the first to evolve sabre teeth, 48 million years ago. About 40 million years ago, the palaeo-sabres invented sabres and used them successfully until their extinction about 6 million years ago. In South America, the marsupial *Thylacosmilus*, the Pouch-knife, had sabres too (see page 15). The extinctions of both palaeo-sabres and Pouch-knives coincided with their encounters with the last family to evolve sabres, the true cats.

Large, sabre-toothed cats could tackle the mega-herbivores, huge rhino-like ungulates (hooved mammals) whose leathery hides were almost impervious to conventional attack. One of the best known is *Smilodon*, which arose in North America about two million years ago. Roughly the size of a Snow Leopard, *Smilodon* had a stubby tail and very powerful forequarters adapted to grappling with prey. Its thick muscular neck ended in a relatively large head with small eyes, sloping forehead and a mouth that bore sabres 18 centimetres (7 inches) along their outer curvature. People speculated that sabre teeth were used like ice-axes for climbing trees, or like a Walrus' tusks for grubbing for food, or to pierce the prey's skull. It was also suggested that they were used to prise apart backbones as small cats do today. Today Jaguars, which relative to their size have the most powerful bite of any big cat, use their canines to pierce the carapace and armour of turtles, tortoises and cayman that make up to a third of their prey in Peruvian rainforest. They are also the only big cats routinely to pierce their prey's skull. However, calculations reveal that while the stiletto-like canines of *Smilodon* would readily skewer flesh, they would shatter on impact with bone. *Smilodon*'s jaw could open to an amazing gape of 90 degrees, compared with 70 degrees for modern cats. Although *Smilodon*'s jaw muscles were relatively enormous it seems unlikely that, at maximum gape, they could have powered two thick sabres through leathery hide and sinew, and nor could the neck muscles have hammered the head and teeth down hard enough to stab the prey.

A clue to how sabres might really function came, unexpectedly, from the mouth of a reptile on the tropical island of Komodo. The 3-metre (10-foot) long Komodo Dragon uses a battery of sabres to cut flesh as it rips into goats. Judging by this working model, *Smilodon* used its sabre teeth as blades rather than stilettos, perhaps butting its lower canines into the prey's flanks or throat to provide a pivot against which the muscles of the head could push down the sabres. As it closed its jaws on prey, the sabres, which had serrated rear cutting edges, sliced backwards and outwards through the hide. The bite might have been used to tear open the prey's belly, severing a rich supply of blood vessels with minimal risk of teeth hitting bone. Alternatively, a sabre-toothed cat might have delivered a bite to the throat. One argument against the throat bite is that it might have involved *Smilodon* opening its jaws so wide agape that the cat would have been unable to see its target. It would thus have risked broken teeth through an accidental strike on bone. However, an unsighted bite might not have been a problem for *Smilodon* if it, like modern cats, could 'feel' with its teeth.

Biologists have puzzled over how Smilodon *used its massive canine teeth*

One idea was that the Smilodon *plunged its sabres into the neck of prey like* Litopterns *(see sequence above left). Although this is how small cats use their canine teeth, biologists have discovered that* Smilodon *could not have hunted this way because its teeth would have shattered on impact (left)*

Smilodon *probably killed mega-herbivores by attacking them from below*

The soft-tissues of the throat pose no danger of breakage for the sabres

Smilodon's enormously wide gape probably allowed it to kill its prey by tearing out their throats, using its sabres as blades rather than as stilettos

Another puzzle was how *Smilodon* managed to eat with a mouthful of sabres. Oddly enough, this may have been resolved by some film-makers who glued fake sabres on to modern Lions in an attempt to reconstruct a sabre-tooth kill. Apparently the Lions ate without difficulty. Sabre teeth were probably also useful for slicing up large corpses for transport. In Friesenhahm, Texas, the den of a fossil sabre-tooth called *Homotherium* is full of the severed legs of infant mammoths, probably carried home to sabre-toothed kittens by their mother. This distinguishes it from modern big cats which take their cubs on hunting expeditions, rather than dragging prey back to a den. Tiger cubs, for example, accompany their mother from two months old. *Homotherium* had especially long carnassial scissors, sharpened to a cutting edge at the rear, which would have easily sheared through elephantine carcasses.

The sabre-tooth design was dominant among large cats at the end of the Miocene, between five and six million years ago, when the world's climate changed in ways that revolutionised the lives of most Carnivore familes (see p. 121). Crucially, the new climate accelerated the replacement of forest by scrub and savannas. The opening up of the plains caused an explosion in the evolution of rodents and also led, four million years ago, to a profusion of antelope and gazelle. These fleet-footed ungulates could evade the ponderous sabre-tooths, but a new lineage of swifter,

Early cats, like Homotherium *(above), were specialists in ambush tactics*

The African Serval (opposite) specialises in hunting grassland rodents and is a small member of the pantherine lineage

more agile cats, the pantherines, rose to the challenge. Their prey had soft hides, so the pantherine cats neither needed sabre teeth, nor could afford their unwieldiness. Today their descendants include the larger members of the cat family. Amongst the oldest are the Puma, which exceeds 100 kilograms (220 pounds) and the Asian Golden Cat which, at 10 kilograms (22 pounds), can tackle small deer. Not all are big: the Leopard Cat is only 5 kilograms (11 pounds), the Rusty-spotted cat is half that size and the Black-footed Cat weighs only 1 kilogram (2.2 pounds). The pantherine lineage has produced three times as many surviving species as either the ocelot or wildcat lineage, and has spread into their ranges. The Puma, Jaguar and the short-legged Jaguarundi or Otter-cat, for example, are all pantherines inhabiting South America, the last stronghold of the ocelot lineage. Two pantherines, the Leopard and the Puma, are medium-sized, all-purpose cats with great jumping ability and immense ranges in the Old and New Worlds respectively.

Most of the pantherines, from Serval to Tiger, share external marks – a white spot behind each ear – and it seems likely that the very first pantherine had such marks six millions years ago. A common speculation is that the spots function as 'follow me' signals to kittens, well out of sight of prey. However, this idea is weakened because the spots occur in both sexes whereas male cats have almost nothing to do with their kittens. Perhaps the spots are a coverable badge, to be

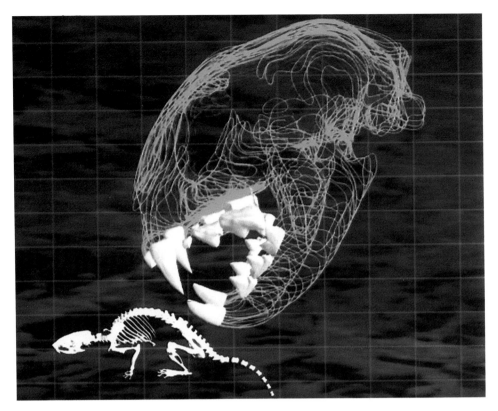

The Serval's short jaw and sharp canine teeth are specialised for stabbing its prey

flashed as a signal of aggression, or perhaps they appear like eyes in the back of the head to an enemy creeping up on a cat. If so, this principle has been borrowed by people in India's Sundarban forest who wear facial masks on the backs of their heads to prevent man-eating Tigers from stalking them.

The new lineage relied on power and speed in the hunt for prey. The African Serval has a long-legged, almost dog-like build, which allows it to pounce over tall grasses for rodents, leap for birds and sprint after small ungulates. The Caracal, on steppes from Turkestan to the Cape of Good Hope, is even more mobile. But the pantherine that, more than any other, took up the challenge of sprinting after the avant-garde antelope of the new grasslands was the Cheetah.

The Cheetah, whose closest living relative is the Puma, broke the cat mould to become more dog-like as a sprinter able to run after prey in the open. Indeed, it is the fastest animal on four legs ever to have lived, its acceleration comparing favourably with that of a high-powered sports car. The Cheetah's flexible spine, long legs and counterbalancing tail all contribute to its explosive sprint. They allow it to bend in action like a looping caterpillar, the body fully extended as the forelimbs are thrust forward and flexed as the hind limbs are brought forward. At a slow gallop, this action adds about 76 centimetres (30 inches) to the length of stride that

The nerves of the Serval's stabbing canines allow it to feel for the gaps between the prey's vertebrae

might be expected from a Cheetah-sized cat lacking such adaptations. As the Cheetah accelerates, the time taken for a complete cycle remains virtually constant at just under a third of a second but the distance covered per cycle increases.

The designers of sports shoes would do well to study Cheetah feet, if they have not already done so. The Cheetah's toe pads are uncommonly hard, and pointed at the front, while the central paw pads bear a pair of longitudinal ridges, like anti-skid tyre treads. Unique among cats, the Cheetah and, puzzlingly, Flat-heated Cats, have blunt claws. The claws are almost as retractile as those of other cats, but are shorter, straighter and have only a short protective sheath, and so act like spikes on running shoes. In contrast, the dew claws on the inside of the forelegs are exceptionally sharp and curved, and function as grappling irons when the cat slaps its fleeing prey. The Cheetah's canines are relatively small, so it kills by strangling its prey with a choking bite to the throat. Built for speed, the Cheetah's head is small, and the roots that used to anchor the upper canines of its ancestors have been sacrificed to free space in the nasal passages. This is necessary to ensure sufficient air flow when the animal is running and or choking prey. Even so, after making a kill, a Cheetah may have to rest for up to half an hour before feeding. Its chewing apparatus is too weak to cope with bone or hide and the skeleton, cleaned of meat, is left intact with the skin lying in a neat pile to one side.

On the Serengeti plains in Tanzania, Cheetah cubs are born throughout the year. Littermates flaunt their kinship on their tails, which bear closely matching patterns of rings that are even more similar to each other's than to their mother's. This is the opposite of what would be expected if the patterns were passed on through genes and has led to the suggestion that the tail banding is determined in the mother's womb. The cubs are weaned at 14 weeks, but may remain dependent on their mother for 13 to 20 months. She brings back small prey alive for them to practise on. Young females leave home before their second birthday, often staying together for about six months. Then they separate and establish 800 square kilometre (300 square mile) ranges, within which each will avoid her own kind and track the game migrations. Young males are chased off their mother's range by adult males, but most remain in company with their brothers, if they have any.

Almost 20 per cent of male Cheetahs live in trios, twice as many in duos, and the remainder alone. Some hold territories while others are nomadic. Coalitions fight as a team and are far more likely than single males to secure and retain a territory. One epic encounter between three collaborating territorial males attacking three nomadic intruders ended with one intruder dead, one injured and one driven off, whereas the defenders were unharmed except for one bloody lip. As well as often being found dead or battle-torn, nomads tend to be in poor condition and rarely mate. Coalitions of males holding a territory, on the other hand, have females drifting on to their property during the wet season. So, male Cheetahs can increase their reproductive success through collaboration. Nearly all coalitions of males involve brothers, and this may be why mother Cheetahs spend more time hunting and provide more meals when they have litters of two or three sons than when they have only one son and several daughters. Mothers of son-biased litters also spend less time feeding at carcasses, possibly stinting themselves while their sons feed. The result is that males from son-rich litters grow up to be bigger than single sons. This is likely to increase their number of offspring still more because larger males retain their territories for longer.

Through strength of numbers, coalitions can claim the best areas, the territories of trios supporting consistently higher prey populations than those occupied by single male Cheetahs. Where antelope are plentiful, there is sufficient food for a male and one or two companions. The females, however, prey mainly on gazelle, and one gazelle is scarcely big enough to share with the cubs that accompany each female for eighteen months. Thereafter the adolescents roam in her territory as a gang for several months more before dispersing and so her territory must provide resources to cater for them too. Furthermore, female Cheetahs would probably gain

A male cheetah scent marks with a fine spray of urine. Males in brotherly coalitions often scent mark together

little from cooperating in defence since two or three adult female Cheetah would be easily overwhelmed by competing Lions or Spotted Hyaena. So, although the home ranges of related females may overlap, they operate alone.

The earliest Cheetah-like fossils occur in Europe and North America, and date back 2.5 million years. With a sprinter's lightweight build, the Cheetah could only tackle small antelope. So, when it evolved, the sabre-tooths were still able to monopolise the role of heavyweight predators specialising in mega-herbivores. Their nearest rivals were middle-weight and light-weight pantherines, such as the Clouded Leopard and Serval. However, large, agile prey, like wildebeest, oryx and zebra, were too fast for the sabre-tooths and too strong for these pantherines. Light-heavyweights, combining strength and agility, were needed; and over 2 million years ago an unknown ancestor grasped this opportunity. One of its descendants became the Lynx, the stockily built powerhouse predator of the northern snows. Another, the Marbled Cat, opted for obscurity in the trees of Borneo. A third gave rise to the five big cats: the Jaguar, the Leopard, the Snow Leopard, and lastly the sister species, the Lion and Tiger. Lion-like cats emerged 1.8 million years ago, and true Lions, 25 per cent bigger than today's version, arose 600,000 years ago. They spanned Europe, Asia and Alaska, and penetrated North America in the form of *Panthera atrox*, which may even have been the same species as modern Lions. These Eurasian and American Lions declined when Ice Age climatic fluctuations changed savanna scrub to forest.

For a while the sabre-toothed sumos and the light-heavyweight pantherines lived side by side on their respective prey. Then disaster struck the mega-herbivores. The weather became both colder and less predictable, which seems to have disadvantaged them. Their absence from cave art suggests that most sabre-tooths were extinct 35,000 years ago and before the final collapse of their prey. Perhaps the evolution of humans was also involved. At first, early humans probably scavenged from the kills of large cats but then they started hunting with weapons. In time, along with Lions, Leopards and Spotted Hyaenas, human hunters migrated from Africa to Eurasia and around the world. Although scientists remain divided on the evidence, the extinction of mega-herbivores and sabre-tooths followed suspiciously close in our wake.

In the cave lairs of Leopards and sabre-tooths, the punctured skulls of early humans, known as australopithecines, indicate that initially our family fared badly in encounters with big cats. Then our genus *Homo* arose and turned the tables, evicting the big cats from the very caves in which they had eaten our ancestors. One can imagine *Smilodon* facing a band of early humans, perhaps clad in *Smilodon* robes, whooping and lunging as they closed in, spears bristling. As their prey dwindled and competition with humans for the remainder became fiercer, the big sabre-tooths declined. Again they illustrate the evolutionary peril of specialisation. They were marvellously successful at killing huge, thick-skinned herbivores, but when these prey disappeared they became obsolete.

For unknown reasons the late Pleistocene was tumultuous for large Carnivores everywhere: the Cheetah's previously worldwide distribution contracted dramatically, and 9400 years ago the last known *Smilodon* died. The demise of the many small sabre-toothed cats, some no bigger than a modern Lynx, is also hard to explain. Perhaps they were ousted by the Lynx-like pantherine cats whose more robust, less breakable dagger teeth provided a competitive edge. Only one modern pantherine, the 20-kilogram (44-pound) Clouded Leopard, has canines almost as long as a sabre-tooth's. It apparently uses its 44-millimetre (1.7-inch) canines to kill wild pigs.

With the demise of the mega-herbivores, the antelope, zebra and other fleet-footed grazers flourished. Had they been the only ambush practitioners on the scene, the large sabre-toothed cats might have been able to survive by feeding on old or weak animals among the new prey, but a Lion can sprint at 60 kilometres (nearly 40 miles) per hour, despite weighing 200 kilograms (440 pounds). Loss of the mega-herbivores had brought the sabre-tooths to their knees, and then competition with the pantherines finished them off. All modern big cats are faster, more agile and, in consequence, less heavily armed than their sabre-toothed equivalents. Like the sabre-tooths, they cannot smash the backbones of large prey buried deep in flesh

The Lion's skull illustrates that cats have huge carnassial scissor teeth, but have lost their rearward molars. Armed with a foreshortened muzzle, they can deliver a massively powerful stabbing bite

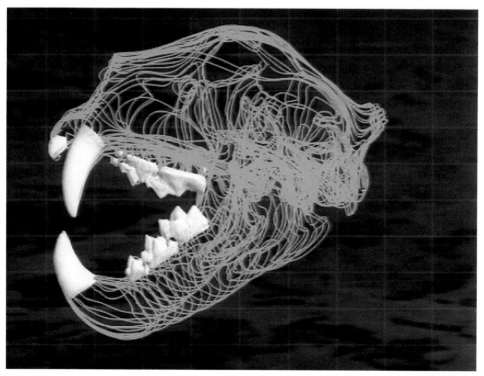

and protected by horns. Tigers kill medium-sized Axis Deer by smashing their vertebrae, but they tackle the larger Sambar Deer from below. Unlike the sabre-tooths, they do not tear out the throat but throttle their prey with a bite that often does not even break the skin. The largest pantherines, such as Lions and Tigers, use their conical canine teeth like anchors, clamping the prey's windpipe in a vice-like grip or engulfing the prey's muzzle in their jaws.

The advent of the Lions brought a new social organisation to the cat family: the pride. Like that of most other Carnivores, Lion society differs from place to place. Most is known about the Lions of the Serengeti which live in prides consisting of between two and eighteen (average six) related females and their dependent cubs, plus one to seven males which may be related to one another but not to the females. Although the pride occupies a territory, its members are rarely found all together. It operates as a number of smaller, ephemeral groups which form and disintegrate every few days. Most Serengeti lionesses spend their entire lives in their parents' pride, but about a quarter of them may leave home, their departure triggered by a new generation of cubs or a take-over by a new coalition of males. While they may disperse with companions or alone, once on the road they almost never join forces with unrelated lionesses. They can generate company only by producing daughters to form a new pride.

Male Lions disperse when they reach sexual maturity at 3 years old. They leave home in a gang which seeks to gain access to females by deposing the resident males of a pride. In northern Tanzania, bachelor gangs of five or more Lions are apparently always composed of relatives, but smaller coalitions often include unrelated travelling companions that have teamed up on the road. In almost half of pride take-overs, at least one member of the triumphant coalition is unrelated to its companions. The wandering bachelors are beset by tribulations, and many perish. Larger coalitions have a better chance of winning, and thereafter of retaining, a pride of females. A male's dominion within a pride lasts on average three years (maximum nine years), and so is usually too brief for father/daughter mating to arise, females becoming sexually mature at around three to four years. If a lioness does come on heat while her father or uncles are still in residence, she may leave home briefly to seek a mate elsewhere.

The related lionesses nurse and guard each others' infants. However, their care does not protect the cubs from the single most important cause of infant mortality amongst Serengeti Lions: infanticide. One third of cub deaths are in the jaws of males taking over their mother's pride. The bereaved mother comes into oestrus on average 134 days after the birth of her cubs, compared with 560 days for females whose cubs survive, or 375 days for females whose cubs die for some other reason. Thus killing infants after a take-over allows incoming males to sire cubs at least eight months earlier than they otherwise would. More cubs survive to their first birthday per mother in medium-sized prides than in larger or smaller ones. Medium-sized prides (five to ten females) are taken over only every four years or so. Females in larger prides are so attractive that they are hard to defend and more prone to take-overs. Small prides of females are often 'annexed' by males from nearby prides, which cannot really spare the time to guard their mistresses effectively. Thus prides of four females suffer take-overs once every fifteen months or so on average.

Competing fiercely for prides of females, male Lions have become mighty but cumbersome. They are fighters, not hunters, easily able to steal food from females and, in open country, dependent upon doing so. Like male Cheetahs, they benefit from allies in the struggle to secure and retain females. The number of cubs a male Lion sires in his lifetime depends on the size of the bachelor coalition he joined. Those operating alone or in twosomes average less than one yearling per lifetime, while those in coalitions of five to seven males each average 3.45 offspring surviving to one year.

So, like the Cheetah mother, a lioness gazing into a crystal ball and wondering how many grand-offspring her son will sire for her will have a good chance of guessing the answer from the number of male youngsters with which he grew up. If he has plenty of male playmates with whom to form a coalition, he is likely to look back years later on a fruitful reproductive career. Females benefit less from belonging to large cohorts. Thus if a mother could predict when she and her pride-

Following pride take-overs, incoming male Lions kill the cubs sired by their predecessors

mates were going to produce a large cohort of cubs, she would do well to choose that time to bear sons – and this is precisely what lionesses do. A lioness might expect there to be a lot of cubs around either when several of her companions whelp synchronously, as in the aftermath of pride take-overs, or when she herself is pregnant with a large litter. In general, lionesses produce equal numbers of sons and daughters, but the cubs born after a take-over are predominantly male, as are those born in larger litters.

Within a pride, males may be surprisingly unpossessive, and sometimes several mate with the same lioness in quick succession. This used to be explained on the grounds that they were likely to be full brothers, and the difference between producing sons and nephews was insufficient to make it worthwhile jeopardising the coalition by squabbling over females. However, closer inspection of Lions' orgiastic affairs has revealed that there is rivalry between the consorts. Males 'claim' oestrous females by guarding them for up to two days before and up to seven days after the three- to five-day oestrous period, during which they mate at the awesome rate of once every 20 minutes. A male can usually guard only one female at a time and females often come into heat synchronously, so members of a small coalition

Male Lions may mate once every 20 minutes during the 5 days of the female's oestrus

have little choice but to share females fairly equitably. In larger coalitions, though, there is a higher chance of there being insufficient females to go around and thus a higher probability of weedier males losing out. This may explain why 75 per cent of duos or trios contain unrelated males, whereas larger coalitions are invariably kin. Members of smaller coalitions need all the help they can get, related or otherwise, whereas members of bigger coalitions may lose more by accepting an unrelated rival for matings than they gain in corporate strength from his presence.

At first it was thought that Lions live in prides because their large, fleet prey can be hunted most efficiently by a group. Certainly, Lions (or more accurately, lionesses) do hunt in groups and end up feeding on the spoils in parties that average between three and eight. However, the link between large prey and living in groups is far from straightforward. For example, male Cheetah fight, but do not hunt cooperatively, in coalitions and a Lion pride seldom hunts as one unit, but splits into temporary hunting parties. Like lionesses, Leopards generally take prey that are slightly heavier than themselves and female Pumas routinely kill prey two or three times their own weight, yet neither Leopards nor Pumas hunt cooperatively. On the other hand, lionesses have been known to take prey more than seven times their own

The big cats generally kill larger prey by suffocation

weight, which is almost twice the ratio for the leopardess's largest prey, and a bit bigger than the female Puma's. However, such massive kills may be rarer episodes in a Puma's life than a lioness's. In addition, it has proven intriguingly difficult to establish whether cooperative hunting does help Lions to kill.

In the Serengeti, a single Lion hunting Thomson's Gazelle is successful on 15 per cent of chases, while two have a capture efficiency of 31 per cent. This efficiency is not increased further, even when up to eight Lions are involved in the chase. When wildebeest or zebra are the prey, a lone Lion is successful on 15 per cent of chases, and a group of six to eight are successful on 43 per cent of occasions. However, such increases in hunting success will often be more than cancelled out by dividing the spoils between collaborators. A further complication is that if hunting in company is more advantageous than hunting alone, a solitary Lion may try to join a group even when doing so is disadvantageous to its existing members.

Outside the Serengeti, it has been easier to demonstrate an advantage to cooperative hunting. In the Kalahari, Lions reputedly hunt porcupine in pairs: one Lion goes for the sharp end, the other at the head; it is hard to imagine how each could cope alone. In the Etosha sands of Namibia, Lions are progressively more successful when hunting cooperatively in prides of up to seven, whereas solitary females are very unsuccessful. Within Namibian prides, each Lion may have a specialist role in the hunting formation, and the kill is always made by one individual. A crucial difference between the Etosha and Serengeti Lions is that the former live in such sparse cover that they need to cooperate while encircling prey.

Even in the Serengeti there are possible advantages to collaborating in the hunt. The Lions' aim may not be to get as much food as possible from each hunt, but to minimise the risk of a run of bad luck. If Lions hunt for Thomson's Gazelle three times a day, then a solitary Lion, successful on 15 per cent of occasions, will on average obtain food every other day. A Lion hunting in a pair, successful on 30 per cent of attempts, will obtain half as much food after each hunt, but will do so twice as often. Thus hunting together would provide greater security, giving each a better chance of obtaining the necessary minimum food daily. Furthermore, hunting in a larger party keeps more options open. Although a single lioness can win a struggle with a buffalo, a hunting party would be much better able to capitalise on a chance encounter with such formidable prey.

Strength of numbers would certainly be an advantage in the struggle against scavengers, such as Spotted Hyaenas, which can make retaining a kill more difficult than killing it in the first place. Big prey cannot be gulped down by a single hunter at one sitting. Leopards solve the problem by dragging their medium-sized prey to a tree-top larder. Lion kills, and Lions themselves, are too big and heavy to take refuge in trees, and even the most formidable Lion can be driven from a kill by a gang of Spotted Hyaenas. So the cost of having to share the booty with companions may be more than offset by the benefit of their support in driving off thieves.

Teamwork not only increases the chances of defending kills but also of pilfering them, and Serengeti Lions scavenge as much as they hunt.

While many factors now fashion Lion society, the origin of their group living may lie in their specialisation on large, open-country herbivores that congregate in grazing herds and at sites such as water-holes and salt licks. Perhaps a solitary, Lynx-like ancestral Lion set up the smallest territory that provided sufficient water-holes and other hunting spots to ensure that she always had a good chance of making a kill. In such a territory, there would generally be food for others as well, because one antelope is big enough to be shared and they travel in herds. So it would have cost the original, solitary ancestor little or nothing to tolerate the presence of, say, her adult daughters. Cohabiting within a territory opens the door to operating as a team if it is advantageous to do so. Acting as guards and wet nurses to each other's cubs would also pay dividends. If the team hunted together they could better have colonised open country and repelled hyaenas. Once female groups evolved and created the impetus for male coalitions, it would become advantageous for adolescent males to be companionable. Such a system then takes on a new momentum, so whatever the origins of their society, nowadays females are social because males are so, and males are social because the females are. The advantages of belonging to a larger club may have become so great that it paid either or both sexes to cooperate in expanding their territory to squeeze in more companions. This could have created a vicious circle in which Lion hunting groups grew to combat scavengers and repel expansionist neighbours more effectively.

While their prey, habitat and rivals allowed – or even compelled – Lions to live in prides, most other big pantherines live scaled-up versions of the exclusive territorial polygyny typical of small cats. For example, female Tigers establish territories and begin breeding between three and four years of age. Thereafter, they can produce litters of three cubs at 20-month intervals for an average of six years. Females may settle in a territory adjoining that of their mother, or she may relinquish part of her territory to a daughter. Some daughters even evict an elderly mother. The end result is that cliques of neighbouring tigresses may be as closely related as lionesses in a pride. Males occupy much larger territories. One, in Nepal's Chitwan Park, maintained exclusive access to seven females over four years during which he fathered 27 offspring that survived to dispersal. Females give males warning that they are coming on heat, which they do throughout the year, by scent-marking and calling. When two neighbouring territorial males meet at an oestrous female they fight, and the winner may take over the loser's territory. At 250 kilograms (550 pounds), males are 50 per cent larger than females and first take over a territory, by defeating the occupant, at around five years of age, and expand it by defeating their neighbours. A cost of this fighting is that the average reproductive life of a male is scarcely three years. The principal cause of cub death is infanticide associated with male take-overs. When the patriarch dominating seven females died, the

ensuing territorial reshuffle caused cub survival to plummet from 90 to 33 per cent.

The Tiger's system in Chitwan causes profound differences in the breeding of the two sexes. Although females and males produced similar numbers of surviving cubs (4.5 and 5.8 on average per lifetime respectively), there was an immense variation in success for individual males. Tigresses produced between nought and twelve survivors in reproductive lives of up to 12 years; yet during the six years in which the most successful male produced 27 surviving cubs, two of his neighbours failed to retain a place in the territorial system and produced none.

Although male Tigers do not tend their cubs, they do coexist harmoniously with their harem of tigresses and their offspring, and may relax and feed with them. Like other big cats, tigresses can hardly be termed solitary, because they cohabit continuously with cubs, a two-year-old generation dispersing when the next litter starts to accompany the mother. The amount by which the females' ranges overlap differs widely from one species of big pantherine to another: female Jaguar, Snow Leopard and Puma may overlap widely, whereas female Leopards rarely overlap. Males' ranges are almost invariably larger and may be exclusive, as in Puma or Leopard, or overlapping as in Snow Leopard. Perhaps whether they overlap depends on the feasibility of defending females. Some Sri Lankan Leopards seem to live in pairs. However, it is hard to see how male Kalahari Leopards could defend their 400 kilometre square (155 square mile) ranges, but easy for Chitwan males to monopolise females within a territory of 10 square kilometres (4 square miles). Similarly, tigress territories in prey-rich Chitwan average 20 square kilometres (8 square miles), whereas those of Siberian tigresses average 250 square kilometres (96 square miles) so males may roam promiscuously their 1000 square kilometre (386 square mile) ranges. One adult male Canadian Lynx is recorded as roaming 783 square kilometres (302 square miles), visiting two widely separated females during one year. Where cats do cohabit they may avoid each others' current hunting patches by monitoring scent marks.

Tiger, Jaguar, Leopard and, especially, Puma at least sometimes kill prey that are, relatively, just as large as those of Lions. The luxury of dense cover to lay ambushes may explain why these big cats do not need a pride to tackle big prey, just as its absence explains why cooperative hunting is so important to Etosha Lions. Tigers stalk to within 20 metres (22 yards) before charging. These and other big cats also face less competition than Lions. Leopards can retreat with their prey to the safety of trees. Snow Leopards can out-manoeuvre wolves on crags. Perhaps in the Americas Pumas and Jaguars were big enough to stave off scavengers such as Dire Wolves – certainly they had no Spotted Hyaenas to contend with. Pumas cover a kill with leaves and may guard it successfully for 19 days. In the absence of serious rivals such as Spotted Hyaenas, Tigers eat their large prey at leisure,

This Sumatran Tigress is part of a typical cat society in which one male monopolises several females

gorging on 35 kilograms (80 pounds) a night, rather than gulping it in company. Yet the Gir Lions of India similarly face no large scavengers and, furthermore, inhabit forest. Although the males are less dependent on females for food and spend more time away from the family than do their African counterparts, the females still form prides. Perhaps they simply carry with them the baggage of history, having evolved from a pride-living ancestor that had presumably travelled out of Africa 500,000 years ago.

Researchers have speculated on the societies of the prehistoric sabre-tooths, particularly *Smilodon*, which was a contemporary of the modern big cats until its extinction less than 10,000 years ago. Drawing on fossil evidence and the lifestyles of modern Carnivores, researchers have reached diametrically opposed conclusions about *Smilodon*'s social life. The 'sociable school' imagines it living in prides, rather like Lions. The 'solitary school' pictures *Smilodon* as slow-witted and heavily-armed, using its teeth in single-handed defence of kills from piratical Dire Wolves. They argue that sociality takes brainwork, and assume that more intelligent species have relatively bigger brains. This assumption may be unwarranted but, in the dog family at least, pack-hunting species have relatively bigger brains than solitary hunters. Relative to its body size, *Smilodon*'s brain was smaller than that of a modern solitary cat such as the Puma, Jaguar and Leopard, and much smaller than that of the sociable Lion.

Fossil *Smilodon* are often damaged, with broken sabre-tooth tips lodged in old wounds. Sabre teeth were precision instruments, and snapping one while fighting would be disastrous. The 'solitary school' interprets these wounds as evidence that *Smilodon* was quarrelsome and hence solitary. However, solitary modern Pumas and Lynx do not seem to scrap more than sociable Lions and Cheetah.

In the Rancho La Brea tar pits of California, the remains of between 1000 and 2000 *Smilodon* are embalmed in awful, glutinous death. Alongside them some 600 large herbivores are glued to their deathbed in the tar. The 'sociable school' uses this as evidence that *Smilodon* hunted in prides, concluding that on average two or three, and as many as eight, sabre-toothed cats were lured to each entrapped large herbivore. However, the sabre-tooths could have just blundered in, as the herbivores did, or been drawn there individually by smaller prey struggling in the asphalt pools. Furthermore, some contemporary Carnivores, like jackals, foxes, and hyaenas, congregate at carcasses smelt from prodigious distances.

Many of the tar-pit *Smilodon* had long-standing deformities resulting from disease or injury. Some had well-healed wounds, showing that the injured animals had survived with disabilities that should have proved fatal if they had been left to their own devices. Amongst modern sociable Carnivores group members may nurse a sick invalid back to health. So the survival of these crippled *Smilodon* may be

A short face and piercing canine fangs are hallmarks of the Leopard's strictly carnivorous lifestyle

evidence of their sociality. Alternatively, their heavy representation at the tar pits may indicate that they were forced by hunger into desperate scavenging.

Perhaps the most promising route to a conclusion about *Smilodon*'s social life is via its prey. *Smilodon* lived on megaherbivores. Would *Smilodon* hunting these prey, and defending their kills, have been individually better off in company or alone? Dense cover might have allowed *Smilodon* to have ambushed single-handedly. If modern Jaguar and Puma have not been forced into prides by the need to repel scavengers, perhaps *Smilodon* could also have managed alone. Once killed one megaherbivore might feed a sabre-tooth for weeks, and at least in the warmer part of their range the meat might spoil fast. Considering the booty to be shared, my guess is that *Smilodon* hunted in prides.

There is one last chapter in the cat story which involves another sociable felid, the Domestic Cat. It arose much more recently than the pantherines but it was equally the result of a new food supply: the hand-outs and rodent-infested grain stores of humans helped to form this wildcat in fashionable clothing. The first records of Domestic Cats come from the ancient village of Deir el Medina, near Luxor in Egypt. Dating back to 150 BC, the village was populated by the artisans who fashioned the magnificent royal tombs nearby. In their own tombs they painted domestic scenes including the Domestic Cat, a symbol of pleasure and fertility.

The first domesticated cats were African Wildcats. Previously the racily built African Wildcat, *Felis lybica*, was classified separately from the stockier European Wildcat, *Felis silvestris*, and the Domestic Cat was known as *Felis catus*. Now all three are lumped together as subspecies of the European Wildcat; and the one that snoozes in front of our hearths is *Felis silvestris catus*.

The ancient Egyptians mummified another species, the Jungle Cat, members of which were kept by priests in a temple built for Queen Hatshepsut around 1470 BC. In the tomb of Beni Hassan, a painting from 2000 BC accurately portrays this species. The Jungle Cat is rumoured to have a more placid temperament than the African Wildcat and this, along with the archaeological evidence, has led to speculation that it, too, played a part in the development of the Domestic Cat. Thousands of mummified cats were found in Egypt, which should have given scientists an excellent record of the Domestic Cat's pedigree. Unfortunately, at the end of the nineteenth century the cat mummies were largely dismissed as uninteresting. In 1889, for example, 19 tonnes of them were shipped to Liverpool and sold as fertiliser. Perhaps modern molecular techniques (see page 8) will confirm whether the Jungle Cat contributed to the Domestic Cat's ancestry.

Domestic Cats are often portrayed as solitary but, in fact, many are highly sociable, in marked contrast to their wild ancestors. Radio-tracking African Wildcats in Saudi Arabia has revealed that, like European Wildcats, they generally travel alone. Even though they had access to dead camels and other carrion, the Saudi Wildcats remained solitary, while the local feral Domestic Cats congregated at these dumps. Perhaps the capacity for such togetherness was fostered by domestication, people selecting sociable cats to increase the numbers that patrolled their grain stores.

Where food is plentiful, Domestic Cats have a remarkable social life. The dozens of cats that may populate a farmyard are not merely a haphazard rabble attracted to free food like moths to a lamp. The females are organised according to family ties, with mothers, daughters, granddaughters, aunts and nieces associating more or less amicably, but excluding rivals of other blood lines. Within these cliques breeding females commonly rear their kittens in a communal nest, each sharing the task of nursing the other's young. Groups of related females may partition the farmyard, and those monopolising locations with easy access to feeding sites rear kittens much more successfully than those banished to the periphery. Like Lions, tom cats kill their rivals' kittens, and the communal denning of the females may increase their chances of having a guard on duty when an infanticidal male calls.

The similarities between the behaviour of female Domestic Cats and lionesses are striking, but between that of tom cats and male Lions there are some obvious differences. Tom cats never operate as a coalition, and each roams alone the ranges of several female matrilines. Male Lions form alliances and generally monopolise only one pride of lionesses. The distribution of prey probably causes prides to be so spread out that males could not defend several, and the males generally need to stick close to females because they rely upon them for hunting. The prey of Domestic Cats, on the other hand, is small, often locally abundant, and taken as ably by males as by females. Farms are often sufficiently close together for one tom cat to be able to patrol several colonies of females.

In their mating, however, male Lions and Domestic Cats share a tolerance that is unusual in the cat world. As many as six adult farmyard toms may consort with an oestrous female at one time, each taking a turn at mating without so much as an impatient cuff between them. Ten different toms may copulate with a female during a single oestrus, and one male may mate with her up to nine times an hour. The female on heat seems to advertise her availability by travelling beyond her normal range spraying urine, which is normally a male prerogative. The arguments advanced to explain the male Lion's lack of aggression to rivals (see page 66) do not seem to apply to tom cats. Amongst male farm cats, at least, there is no known cooperation, nor any evidence of long-term brotherly alliances.

Generally among mammals, an individual will only escalate a contest to bloodshed if there is a reasonable chance that the rewards will outweigh the costs. In a large colony of cats, such as a farmyard accommodating several adult males and females, a number of factors may combine to minimise the benefit of fighting. First, the difference in the contestants' fighting prowess may be slight, feeble individuals having been thrashed before they got close to an oestrous female. To emerge as overall victor, any male would have to beat several evenly matched rivals, and being the first to enter the fray could also mean being the first to become exhausted.

Second, and exacerbating the first problem, the claws of cats are such fearsome weapons that it is hard to imagine even the victor emerging unscathed from any serious combat. Third, it may be impractical to monopolise several adult females, all likely to come on heat at roughly the same time and widely spread out around the farmyard. Finally if, as would be expected, kitten-killers spare their own off-spring, then kittens of uncertain paternity would be at an advantage and so females may choose to mate with several males. All these factors make it likely that the costs of fighting outweigh the reward measured in additional surviving kittens sired. They may, of course, find other ways of competing. For example, the order of mating may be crucial, the first or last to mate gaining a preponderance of the conceptions. Until genetic studies reveal the paternity of kittens, one can only speculate on the courtship of the farmyard lions.

While the societies of many species of cat have been coarsely categorised as solitary or sociable, polygynous or promiscuous, there has been little study of relationships between individuals in these systems. Social ties within groups have only been explored for one subspecies, the Domestic Cat. Amongst farm cats, a social fabric emerges through the asymmetries in aggression and facial rubbing. Females rarely initiate aggression to adult males, whereas males frequently mete out aggression to them. On the other hand, females often approach a male and rub their lips and cheek against his face, whereas males rarely rub on females. Furthermore, the flow of rubbing between particular females is often very one-sided, and a kitten will rub on females irrespective of blood-ties but in proportion to the times it is nursed by them. These asymmetries suggest that rubbing is an indicator of dominance, those in net receipt of rubs being closer to the social centre. So a cat rubbing on its owner may be expressing more than mere friendliness. The puzzle is that it is not clear what extra benefits accrue within matrilines to individuals in receipt of these tokens of social standing.

Just as a kitten rubs on nursing females, so a puppy licks jowls. Both infantile behaviours are incorporated into the repertoire of submissive adults, and both have the capacity to transfer oral scent between the supplicant and dominant. The lips and cheeks of cats are endowed with scent glands, and their saliva too may carry social odours. Thus, when one cat rubs on another it may deposit scent and/or pick it up. Both could be advantageous to the subordinate, one keeping the odour of the dominant on her as a sort of membership badge and the other keeping her smell in the dominant's nose as a receipt for dues paid. Indeed, the amalgam of scents on a cat's cheeks might come to constitute the hallmark of the matriline, itself subsequently smeared on other cats and on objects. Whatever the precise significance of head rubbing, the fact is that all members of the cat family do it. From Lynx to Lion they all bump their heads together in greeting, emphasising again the similarities that have united cats, past and present, big and small, for at least 12 million years. My guess is that greeting *Smilodons* rubbed heads too.

Family Tree: The Cats

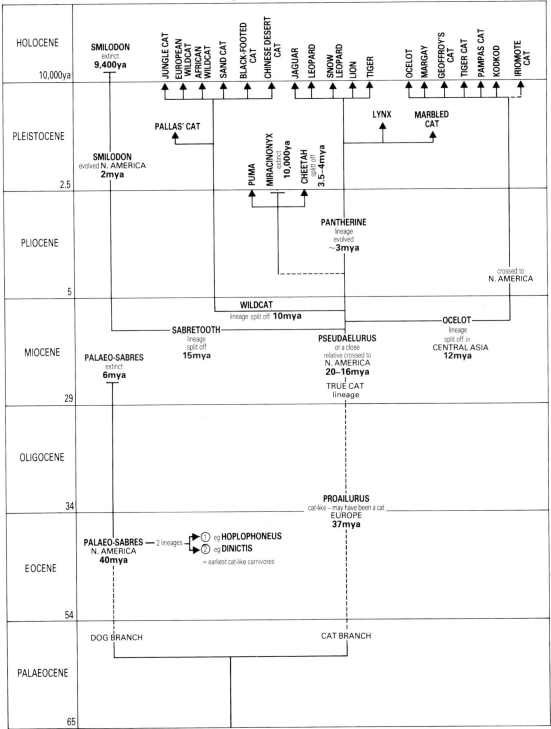

CHAPTER 3

PACKING POWER

When the Carnivores came down from the trees, some became cat-like, depending on stealth and ambush to catch their prey. Others opted for a mobile, generalist lifestyle. It was these that gave rise to the dog family Canidae, which includes wolves, jackals, foxes and their ilk. Members of the dog family tend to be opportunists. The Crab-eating Fox on a nocturnal jaunt may indeed eat a crab, but it may also eat frogs, fowl or fruit as opportunity dictates. The Grey Wolf, weary of pacing after mice, may try its luck with a Moose. Some Coyotes depend heavily on deer, whereas elsewhere they live largely on mice and rabbits, with the result that some Coyotes hunt in packs, while others live in pairs and forage alone. This adaptability allows most species of canids to live under a variety of circumstances.

To discover how the dogs became hunters to rival the cats we must go back in time 40 million years to where their story begins, in North America. The dogs' ancestors hunted the canopy of the jungle that cloaked the northern hemisphere. The first members of the dog family, the dawn dogs, *Hesperocyon*, appeared in North America 35 million years ago. Dawn dogs had long tails and low-slung bodies which doubtless enabled them to slip easily beneath the forest undergrowth as they trotted on, digging at a rotten log, scratching out beetle grubs, pausing beneath a large tree to eat windfall fruits, perhaps even climbing to reach fresher fare.

While their cat-like contemporary, the palaeo-sabre *Hoplophoneus* (see page 43), lay in ambush, wrestled a sizeable prey to the ground and delivered a stabbing bite, the dawn dogs could only look on. During the hours through which *Hoplophoneus* had stalked and ambushed, the dawn dogs would have been travelling on slender legs whose efficiency was bought with diminished climbing and grappling abilities. Being Carnivores, they had scissor teeth, but they cut with a blunter edge, and were backed by broad, spiky-cusped molar teeth designed to crunch as well as cut. *Hesperocyon*'s full complement of teeth required a long muzzle which weakened the

Some early vulpavines of the dog-branch came down to earth and, like Hesperocyon *(above right), adopted an omnivorous life-style.* Hesperocyon, *the dawn dog, would probably have been intimidated by its cat-like contemporary, the palaeo-sabre* Hoplophoneus

forces that enabled the short-faced *Hoplophoneus* to plunge its sabre-teeth deep into large prey. So while *Hoplophoneus* filleted its feast with carnassials that sliced easily through flesh, *Hesperocyon* chewed snacks such as beetles, a small mammal, an egg, a handful of grubs and some fruit.

The dawn dogs' diet of titbits would not have provided much scope for sharing a carcass, nor for pack hunting. But it would have been well worth defending an area from rivals that might empty the larder. Furthermore, two pairs of eyes are better than one. A pair of dawn dogs could have fed side by side on a fruiting bush, and there would have been sufficient beetles and grubs for a pair of them keeping loosely in company as they travelled a territory large enough to sustain them both. Perhaps the omnivorous diet that went with generalist teeth, long muzzles and nimble build also encouraged the monogamous family life that many of their descendants follow today.

The beautiful Grey Fox of the southern USA and Central America is the oldest surviving dog, having evolved between six and nine million years ago, and it lives today much like the dawn dogs once did. With dainty steps, territorial pairs move nimbly through deciduous woodland and deserted farmland, searching for insects, fruits, carrion and small mammals up to the size of rabbits. Bearing a flamboyant fluffy tail, the Grey Fox is as delicate in appearance as in movement. Its long, brindled fur merges with pastel buffs and oranges on its neck, legs, flanks and cheeks, and the face and underside are patterned in white. By the standard of today's dogs, it is a good climber, being able to rotate its forelegs to clasp branches or rocks. In its distinctly three-dimensional world, hunting cliff crevice, tree bowl and root hole, a Grey Fox in a hurry moves like a spring-loaded powder-puff.

The descendants of the dawn dogs diversified in their tropical home. New species can arise as animals adapt to new circumstances, such as a change in climate, competitors, prey or habitat. This can be seen particularly clearly when individuals of a mainland species colonise an island and adapt to conditions there. In isolation, they may evolve such great differences that they can no longer interbreed with their mainland ancestors. This process is, coincidentally, illustrated by the Grey Foxes off the coast of California today.

The Californian Channel Islands are grouped in two clusters, the northern most of which lies less than 30 kilometres (19 miles) off the mainland. On each of the six islands are miniature replicas of the Grey Foxes found on mainland California. The island dwarfs weigh between 1.3 and 2.8 kilograms (2.9 and 6.2 pounds) while the mainland Greys weigh around 5 kilograms (11 pounds). Living on insects and fruit, perhaps the island foxes became smaller because they were freed from competition with other small predators of the mainland. Scientists wondered whether they were different enough to be considered a separate species and, if so, how long this process

The Grey Fox stems from an early canid lineage, and remains the most adept tree-climber of the family

had taken. By combining fossils, archaeological clues and molecular biology, the foxes' history has been unravelled.

Fossils revealed that mainland Grey Foxes first arrived on the northern islands about 16,500 years ago, at a time when the islands were still connected to each other and split from the mainland by only a 4–5-kilometre (2.5–3-mile) channel. The longer two populations have been separate, the greater will be the difference in their DNA, the substance in which an individual's genes are encoded. This difference is called the 'genetic distance' between them. Blood samples from foxes on each island revealed that the genetic distance between mainland and island foxes is as large as that between, say, a Grey Wolf and a Coyote. So in only 16,500 years the breakaway population had evolved into a new species, the Channel Island Fox. Indeed, the populations on each of the six islands are also significantly different from each other and are on the evolutionary road to becoming separate species.

Flexibility of behaviour and mobility allowed the Grey Fox's distant ancestors, the early dogs, to spread out from their jungle cradle. Twenty million years ago a branch of the dog family known as hyaena-dogs, or borophagines, arose; and they flourished until the late Pliocene three million years ago. Among the last to go were *Borophagus* and *Osteoborus*, which crushed bones using carnassial teeth adapted to a sledgehammer design. Modern hyaenas smash bones with a more forward

Osteoborus was one of the borophagines or hyaena-dogs, a subfamily of the dogs that specialised in bone crunching

Enhydrocyon *was the dog family's attempt at producing a big cat, and was like a modern Jaguar*

premolar, leaving their carnassials as scissors, and so enjoy dual-purpose dentition. Other borophagines were jackal-like, and one called *Epicyon* was a large predator that may have hunted in packs, preempting the evolution of wolf-like true dogs in North America. The true dogs also produced some oddities including *Enhydrocyon*, with a short snout and deep jaws reminiscent of a Jaguar.

The early dogs probably competed with the amphicyonids, members of the one Carnivore family that has become extinct. The amphicyonids are called half-dogs or, confusingly, bear-dogs, although they have nothing to do with either bears or dogs nor with the arctocyonid bear-dogs (see page 16). This dog-like family died out at roughly the same time as wolves and predatory running bears evolved, and pantherine cats crossed from Eurasia to America. Whether any or all of these creatures accelerated the half-dogs' demise about eight million years ago in the face of climatic change is unknown.

Between five and seven million years ago, some true dogs moved north to the Bering landbridge joining North America to Asia and, hence, Europe. Either just before or just after they crossed, an early dog called *Canis davisii* probably gave rise to the two main lineages found today: the fox-like vulpine dogs and the wolf-like lupine dogs. Before this split the early dogs gave rise to the ancestors of today's Far Eastern Raccoon Dog. Raccoon Dogs eat a lot of fruit, gorging in autumn to put on weight that carries them through a period of lethargy unique amongst canids. These and Grey Foxes are the only surviving dogs whose origins predate the split

The Raccoon Dog of East Asia (above) is unique amongst dogs in becoming lethargic in winter

The Arctic Fox's amazing fur (opposite) protects it from sub-zero winds

between vulpine and lupine species.

We can guess at the conditions faced by the pioneers as they crossed the landbridge. Just over 30 million years ago the mean annual temperature in Alaska had plummeted to 10°C (50°F), and subsequently it cooled still more. However, there were slight warmings from time to time, and it may be no accident that the dogs crossed during one of these periods. The far north may not have been as brutally cold as it is today but the first dogs to cross to Eurasia would still have faced a gruelling journey.

The modern Arctic Fox's range encircles the North Pole from Alaska to Kamchatka, like a skullcap on the globe. This astoundingly beautiful creature is a monument to the endurance of its kind. The Arctic Fox is a new arrival, a late Ice Age fox, but perhaps the early canids that ventured north were similarly designed to withstand the intense cold. The 3–4-kilogram (6.5–9-pound) Arctic Fox has relatively short ears, legs and tail to minimise heat loss, and its paws are swathed in fluffy mittens of woolly fur. About 70 per cent of the Arctic Fox's fur fibre is fine underfur, compared with 20 per cent in Red Foxes. Its fur is so luxuriant that the Arctic Fox begins to shiver only when the temperature drops to −70°C (−94°F). Indeed, an Arctic Fox caught in a run-of-the-mill blizzard is more likely to be troubled by heatstroke than frostbite.

Arctic Foxes breed on the tundra from 53 degrees north, level with the southern tip of Hudson Bay, to the snowy wastes north of Greenland at 88 degrees north. As the tundra squeezes a whole year's growth into the two-month summer, the Arctic Fox races to breed. It also has to contend with the boom or bust ecology of the far north. A four-year cycle in populations of its lemming and vole prey transforms the banqueting ground of one year to a field of famine in another. In Barrow, Alaska, Brown Lemmings may soar to over 200 per hectare (80 per acre) in one year, then plummet to less than one per hectare (0.4 per acre) in another. Arctic Foxes disappear from Barrow during the troughs, but swarm in at 25 per square kilometre (65 per square mile) during the peaks. They are able to take advantage of the booms by breeding more prolifically than any other canid. The Arctic Fox is equipped with a record-breaking 12 to 16 teats, rivalled only by the African Wild Dog with 12 to 14. Its teats sustain litters averaging more than 11 cubs, the highest recorded being 19 in one litter.

Arctic Foxes change colour with the seasons. During the winter, many have a white cloak which hides them from prey and perhaps from predators, such as wolf and eagle. In spring, the white pelt is shed in moth-eaten, woolly chunks to expose a brown back and limbs, blending to a creamy beige below. On the brownish grassy tundra, this summer fox is well disguised apart from grizzled silver eyebrows which lend a distinguished air quite at odds with impertinent tufts of white hair sprouting from its ears.

The Arctic Fox's adaptable colour scheme exists in two forms – white and blue. While the white or polar form moults from white to brownish-grey, the blue morph swaps a deep brown winter coat, with a metallic bluish sheen, for pebble brown in summer. Whether a fox is white or blue is determined by a gene coding for coat colour, and each fox carries a set of two colour genes. A pure white fox has two white genes and a pure blue fox has two blue genes. If they mate, their offspring will receive a gene from each parent and so will carry one white and one blue gene. All will appear blue because the blue gene is dominant. However, their offspring may be white or blue, depending on their mate's genes, and white and blue cubs may occur in the same litter.

Over the vast expanse of the Arctic Fox's continental range, the white form predominates and fewer than one in a hundred is blue. On shores, where snow does not settle, a white fox sticks out while a beachcombing bluish-brown fox blends discreetly with its surroundings. Not surprisingly, therefore, blue foxes are most common on smaller islands which have a higher ratio of coastline to interior. On the large island of Greenland, for example, the two colours are equally represented, whereas on the smaller Iceland, two-thirds of the population is blue. Blues are in the majority along the coastline of Iceland, where white foxes are always at a disadvantage and only continue to crop up due to immigration from the interior. Inland, blue foxes are better off than whites during summer, being less conspicuous to fox-hunters, whereas white foxes are better off in winter, resulting in a mixture of both. One might expect colour prejudice amongst Arctic Foxes, with mates being selected to produce young of the colour appropriate for local conditions, but Icelandic foxes pair up with no regard to colour.

In Alaska some Arctic Fox populations migrate regularly in search of food. Each autumn, they vacate inland breeding grounds and travel to the coast, returning inland in early spring. Over large expanses of Canada, Fennoscandia and Russia, Arctic Foxes live on a knife-edge honed by the fortunes of lemmings. When lemming populations crash, the foxes face starvation and set forth on mass emigrations. Nobody knows how individual Arctic Foxes fare on such an exodus. Perhaps these modern migrants face similar rigours to those overcome by their forebears trekking west on the journey from North America to Eurasia.

The arrival in Eurasia of dogs, between five and seven million years ago, coincided with the demise of most of the rivals that might have kept them out. It was at roughly this time that the last of the creodonts (see page 23) died out along with the amphicyonid half-dogs and most of the hyaenas which had until then occupied the niche that was later to be exploited by the dogs. The canids cannot be blamed for these extinctions, all of which were probably precipitated by the changing climate. However, the result was that the dogs arrived in the right place at the right time to fill niches left vacant by the demise of potential rivals.

About five million years ago the fox-like vulpine dogs and wolf-like lupine dogs began to diversify. The first offshoot of the lupine dogs in the Old World were the ancestors of the modern Bat-eared Fox of Africa. The earliest vulpine lineage to split off led to the Cape Fox which is now the only vulpine species found south of the Sahara. Another led to the Red Fox which soon spanned Eurasia and early on gave rise to the plains-dwelling Corsac Fox of the Russian steppes. With the frozen Bering crossing behind it, the Red Fox lineage trekked south to sun and sand, adapting to circumstances no less harsh, with the ancestors of the arid-land Swift and Ruppell's Foxes splitting off just after the Corsac.

About three to four million years ago a fox-like ancestor, perhaps an offshoot of the Cape Fox line, found itself facing the climatic changes that produced the first deserts in the Middle East and northern Africa. It produced two desert specialists: the Fennec and the Blanford's Fox. Weighing only 1–1.5 kilograms (2.2–3.3 pounds), the Fennec lives deep in the Sahara. While the Arctic Fox does not shiver until temperatures plummet to −70°C (−94°F), the Fennec starts to tremble with cold at less than 20°C (68°F), and neatly wraps its tail like a stole around its feet. However, it has an amazing record of its own: the Fennec Fox only starts to pant when the temperature exceeds 35°C (95°F), and its jaws open to a full pant only at 38°C (100°F). But when it pants, it really pants, the resting rate of 23 breaths per minute rocketing to as much as 690 breaths per minute.

Fennecs shelter underground during the heat of the day and hunt in the cool of dusk and dawn. They rarely drink, and secure moisture from their food, which includes small rodents, lizards, birds and insects. These foxes are adapted to conserve as much water as possible. A panting Fennec curls its tongue up, so as not to waste even a precious drop of saliva. Its butterfly ears constitute 20 per cent of its body surface and, when the temperature soars, it dilates the blood vessels in its ears and feet. This allows more hot blood to flow through the extremities, increasing the amount of heat radiated to the outside. If the air temperature climbs higher than its normal body temperature of 38.2°C (100.8°F), the Fennec lets its body heat up to 40.9°C (105.6°F), thus reducing the water it has to 'waste' in sweating. The Fennec also saves energy by having a metabolism that chugs along at only 67 per cent of the rate predicted for such a small animal. Similarly, its heart rate of 118 beats per minute is 40 per cent lower than expected.

In 1981 came the amazing news that a new desert fox had been discovered in Israel. It weighed only 1 kilogram (2.2 pounds) and had large ears, a stupendously fluffy tail and naked footpads. Eventually it was identified as Blanford's Fox, which meant that it was not a new species but an old one in a new place. This sister species to the Fennec had previously been known only from remote hills in Afghanistan and Iran, so the Israeli specimen was at least 1400 kilometres (870 miles) off course. Using their sinuous tail as a flamboyant counterbalance and naked footpads for high-traction grip, Blanford's Foxes bound with feather-light grace up near-vertical surfaces on scree, cliff and crag. They live in pairs, sharing cliff territories of 0.5–2 square kilometres (0.2–0.8 square miles) as they quest for beetles, grasshoppers, spiders and berries, specially relishing capers. Like the Fennec, they rely on such food for moisture as well.

Blanford's Foxes are unusual in never caching food, their prey presumably being too tiny and perishable to merit hoarding. Most other canids, with the possible exception of Bat-eared Foxes, do store food and have excellent memories for their hiding places. This habit constitutes an invaluable insurance policy for opportunistic hunters that never know where their next meal is coming from. Hoarding may be another advantage of the dogs' robust feet and claws which are well adapted to digging hiding places. Their dainty feet make cats poor diggers and, excepting Leopards which hoard up trees, most rarely make caches. Indeed, aside from Spotted Hyaenas which store carcasses under water, food-hoarding is rare in cat-branch Carnivores.

The Blanford's Fox's ability to live on cliffs explains how it can coexist with the Fennec of shifting sands. The south of the Fennec's range also overlaps with that of the Pale Fox of the southern Sahara, and, to the north, with that of the furry-footed Ruppell's Fox. Compared with the tiny Fennec and Blanford's, the Ruppell's is a newcomer to the desert scene, being a miniature version of the Red Fox from

The Blanford's Fox is specialised for cliff-dwelling, with naked footpads for extra traction

which it recently descended. In addition, Red Foxes also penetrate Middle Eastern deserts where they are 3-kilogram (6.6-pound) waifs compared to their European brethren which can weigh over 10 kilograms (22 pounds). Competition between species as similar as Reds and Ruppell's would normally result in the larger ousting the smaller. The general rule is that larger dog species dominate, or even kill, smaller ones. Grey Wolves chase and kill Coyotes; Coyotes kill Kit Foxes; Golden Jackals kill Red Foxes; and Red Foxes kill Arctic and Grey Foxes. However, in some areas smaller foxes are, paradoxically, protected by their size. This is illustrated by the distributions of the Red and Arctic Foxes. The Arctic Fox's range forms a circumpolar ribbon to the north of the Red's, the two species overlapping in the Eurasian and Canadian tundra. Both species are impressively adapted to the cold, and are remarkably similar in everything they do. As a result, when they meet, Red Foxes seem to treat Arctic Foxes as smaller, and thus inferior, copies of themselves and overpower them. However, the heavier Red Foxes need to eat far more and towards the north food becomes too thin on the ground to sustain them. The larger body that allows the Red Fox to bully the Arctic Fox further south gives it an appetite that cannot be satisfied to the north. So, the Red Fox's brute strength sets the southern limit to the Arctic Fox's range, while its hefty appetite sets its own northern limit.

Similarly, the small Ruppell's Fox may thrive in arid landscapes too impoverished for the larger Red Fox. In other areas competition between the two species may be defused by exaggerated differences in the lengths of blades on their lower scissor teeth. These differ more in parts of Israel where Reds and Ruppell's coexist than they do where only one of these species occurs, perhaps allowing them to specialise on different diets. It seems odd that such small adjustments in chewing teeth could allow them to cohabit. However, an equivalent small difference in size separates the canine teeth of Red Foxes from those of the Grey Foxes with which they coexist in the USA. Similarly, Golden Jackals have longer scissor blades in north Africa than they do in the Middle East, where they are in competition with the larger Arabian Wolves. Although the scissor blades of jackals are the same length from Egypt to Morroco, their bodies are much larger in the east than the west. Perhaps this means that they all eat the same range of foods, but it is scarcer to the west. However even quite big differences in size may not defuse competition between some species: the main cause of death for San Joaquim Kit Foxes and Canadian Swift Foxes is predation by Coyotes.

There are also species that diverge in size in some areas, such as the Chilean Grey Fox and Culpaeo of the Andes. These two descended from a common ancestor less than half a million years ago, and differ principally in that the Culpaeo has longer canine teeth. Northern Culpaeos and Greys are both about 70 centimetres (28 inches) long, whereas in southern Chile they are 90 centimetres (35 inches) and 60 centimetres (24 inches) long respectively. The northern Andes are high and the great altitudinal range of habitats provides a diversity of rodent prey, while in the lower southern Andes prey diversity diminishes. Perhaps the size difference between the two species in the south is due to the fact that a less varied habitat and prey forces them into direct competition with each other. Another puzzle is that three species of jackal coexist in east Africa with only slight differences between them. The Black-backed, Golden and Side-striped Jackals arose three to four million years ago. All three may meet at the same carcass. Golden Jackals tend to live on the plains, Black-backs in thorn-scrub and Side-stripes on hillsides. Whether these separations are enough to defuse competition between them, or whether they can co-exist because they are limited by disease or predators rather than food remains an open question.

Wherever they occur, vulpine foxes will eat almost anything they can find, but they do have their specialities. The Red Fox, the largest of modern vulpines, is a specialist mouser. This specialisation has given it a lot in common with small cats, including the ability to pounce on prey with great precision. The fox detects its prey by pinpointing the sounds of its hidden victim. The Red Fox can locate sounds to within 1 degree, and its hearing is especially acute at the low frequencies at which rodents rustle in the vegetation. Then it springs, sailing high above the quarry, beating its tail to steer in mid-air, before plummeting down to land up to 5 metres (16 feet) away, and smack on target.

A number of factors help the Red Foxes achieve their accurate pounce. First, they normally take off at an angle of about 40 degrees, which is close to the theoretical optimum of 45 degrees for maximum distance. They aim lower for short jumps, and much higher, up to 80 degrees, if they need to land with extra force, for example to break through a crust of snow. Second, Red Foxes have relatively longer hind legs than other members of the dog family, and this increases their propulsive force. Third, they are much lighter than other dogs of the same size, being half the weight of a domestic dog or small female Coyote of fox length. Their skeleton is streamlined, the leg bones, for example, being disproportionately slender and weighing 30 per cent less per unit area of bone than expected.

Despite all these adaptations, there are areas where Red Foxes rarely eat mice, and everywhere they retain the opportunism that secured their ancestors' success. For example, in close-cropped pasture they take advantage of the easy access to succulent, crawling fare. There is great precision in a Red Fox's pursuit of earth-worms which, as every fisherman knows, crawl to the surface on still, warm nights. Some worms slither unprotected through the jungle of grass while others retain a hold in the soil, with tails anchored and bristly chaetae well dug in. The fox listens, with head cocked and ears perked, for tell-tale sounds over the din of rustling leaves. Suddenly, the fox freezes, brush poker-stiff and ears flicking. A worm has incautiously rasped its chaetae on the grass and the fox's snout points at the noisy indiscretion. It plunges down into the grass and the worm's head is neatly clamped between the fox's incisors, its tail wedged 'safe' in the soil. The fox avoids snapping its victim by holding the worm momentarily taut and then raising its muzzle in an arc, slowly at first, to draw the worm intact from its sanctuary. Flinging itself into contortions, the earthworm winds around the fox's muzzle, but the captor, its jaws manoeuvring dextrously, slips the animated spaghetti down its throat.

For the first 30 million years of their history, most canids probably lived much like foxes, hunting alone for titbits. This does not mean that they were anti-social, however. Both Red and Arctic Foxes generally forage alone and sometimes live in territorial pairs, but in some habitats they also form loose-knit social groups. These tend to consist of a male and two to five adult females. The females are generally close kin, often a mother and her daughters.

In three Arctic Fox territories that were studied along the shoreline of an Icelandic fjord, a male and two vixens cohabited in each, living by beachcombing for carrion. The territories varied hugely in size, the largest embracing 10.5 kilometres (6.5 miles) of coastline and the smallest 5.4 kilometres (3.4 miles). This may have meant that one group garnered twice as much carrion as another or that the smaller territories had more productive coastline. The deposition of carrion varied with the run of the tide and lie of the land. So to assess productivity it seemed that the researchers would have to measure the grubs, dead fish and other sea life that each group of Arctic Foxes could glean from their beaches. Fortunately, a local farmer, who combed the coastline for driftwood from which to make fenceposts, kept a diary of the yield from each stretch of beach. Since driftwood and fox food were likely to be washed up in the same pattern, it was possible to calibrate territory quality in units of fenceposts! It turned out that, despite their wide difference in size, all three territories yielded almost exactly the same number of posts. Paradoxically, the occupants of the largest territory were worse off than their neighbours because they had a similar number of productive coves but further to travel between them.

This study also provided a clue as to how such groups arise. The shortest stretch of coastline productive enough to guarantee an adequate food supply for a pair of Arctic Foxes will often provide enough food for a companion too. Any dead seals, or even seabirds, washed up would be a feast for several foxes and, at other times, the junior member, generally a yearling daughter, might have to make do with seaweed maggots. This is quite different to the Arctic Foxes on Wrangel Island in Russia. They live on lemmings which are evenly spread across the tundra so a pair of foxes can expand or contract its territory as lemming numbers soar or crash.

Blanford's Foxes preying on crickets in a sun-baked canyon have much in common with Arctic Foxes scavenging on an icy beach. The cliff-face territories of the Blanford's Fox also vary greatly in size, ranging from 50 to 200 hectares (120

The Red Fox now has the widest distribution of any Carnivore, spanning the Northern Hemisphere and much of Australia. Its opportunism epitomises the dog family's evolutionary success

to 500 acres). They concentrate most of their hunting in the fertile floors of deeply cut wadis, rich in insect prey. However each territory, whatever its area, contains roughly the same amount of creekbed. The Blanford's Fox's wadi is, in effect, the Arctic Fox's productive cove. However, in seven Blanford's Fox territories studied, five were occupied by a pair, and in only two did a probable daughter cohabit. The amount of insect prey available to those foxes does not vary much from night to night, so the smallest territory required by a pair will not often support a third fox.

Red Foxes studied on the outskirts of Oxford, on the other hand, were found to live in groups of four or five adults. They feed mainly on food scavenged from bird-tables and compost heaps, and windfall fruit and earthworms. These foods are quite plentiful and the territories are correspondingly small, ranging from 19 to 72 hectares (47 to 178 acres). The foods also vary widely in their availability. Whether a given field is heaving with earthworms or yields none at all will change from night to night, or even hour to hour. A fickle change in the wind will drive the worms underground before their moist bodies become chilled. So, a pair of Red Foxes needs several pastures to ensure that one will be in the lee of the breeze. Yet each pasture will probably provide enough worms to feed several foxes. Similarly, households differ in when they produce scraps: in a given garden foxes might find a bonanza of chicken bones one night, and nothing the next. So each territory has to encompass enough homes to give the foxes an acceptable chance of eating every night. All seven Red Fox territories in the area, despite varying in size, have close to 24 human households within their borders. This magic number is apparently enough to guarantee scraps for a pair and, incidentally, to give two or three of their adult daughters a reasonable living too.

The early wolf-like lupine dogs diversifying in Eurasia were probably generalists like their vulpine contemporaries. They were much like modern Coyotes and, being a bit bigger than foxes, probably relied more heavily on rather larger prey. They doubtless also scavenged at carcasses, as one of their oldest living representatives, the Black-backed Jackal, does today. The carcasses might have supported the members of a family which, by joining forces, improved their chances of holding off other scavengers. Large packs of Canadian Coyotes have greater success in defending prey carcasses from rivals than do smaller packs. Coyotes form packs of up to eight where the carrion include large deer whereas they live in pairs in areas where they feed primarily on mice.

Members of a group may also cooperate occasionally in hunting. Black-backed Jackals and Golden Jackals are three times more successful in hunting gazelle fawns when they operate in pairs than when they hunt alone, one jackal perhaps distracting a mother while the other kills the fawn. Coyotes are quick to cooperate, and in some areas even form remarkable alliances with American Badgers to hunt for ground squirrels. The improbable pair forage side by side as they quarter the brush, helping each other to flush small prey which may be caught by either partner. The alliance is slightly uneasy, because badgers occasionally kill and eat Coyote pups, and a family of three adult Coyotes has been seen killing a badger. Thus in the hunt, as in defence, the early dogs probably teamed up with one of their own species, particularly a family member which they were likely to be on good terms with already.

Joining up with kin also offers opportunities for other types of cooperation, including acting as a nanny. On the grass plains of the Serengeti a Black-backed Jackal soars through the air and then draws its hind legs in tight below as it thuds, forepaws first, on to its victim. Grass rat swinging from its jaws, this bushy-tailed, slender-muzzled dog heads for home. Having nabbed a second rat en route, the male jackal arrives at the den. A female and four stumbling pups greet him. Tails wag in feverish greeting, ears flick back, necks twist, jaws gape and there is much whining. The male drops the two rats and a macabre tug-of-war ensues amongst the pups. It is a typical family scene, except that the male is not father to the pups, nor is the female their mother. They are elder brother and sister to the brood: individuals which have opted to stay at home as nannies rather than strike out on their own.

Amongst Black-backs babysitters join their parents in diligent care of their younger siblings. They hunt avidly to feed them and spend long hours guarding the den, playing with the pups and grooming them. They even risk their lives to defend the brood against marauding Spotted Hyaenas, for which a jackal pup represents an appetising morsel. Dens may be attended by from one to four nannies, which are equally likely to be males or females. The more helpers there are, up to about four, the more pups will survive. A jackal pair raising a litter unaided will often have to leave their pups on their own for up to 40 per cent of the time after the first three weeks of almost constant attention. Nannies mean that the young will seldom be unguarded and will receive more food.

Helpers are not restricted to the lupine lineage. Some Red Foxes and Arctic Foxes have them too, although their nannies are almost invariably female. In one case, a mother Red Fox's injuries prevented her from tending her weanling young. Aided by the father, the adult sisters successfully tended the cubs. In another case, a mother Red Fox died. Her adult daughter, who was barren and had until then occupied a distant corner of the mother's territory and shown no interest in the cubs, immediately moved in and successfully reared them.

Whether to stay at home as a helper or set forth on a perilous journey to seek its own territory is a difficult choice for a newly adult member of a family group. Among Golden and Black-backed Jackals, Bat-eared Foxes, Crab-eating Foxes, African Wild Dogs Foxes and Coyotes, some individuals have left home for months on end only to return to their parental home as assiduous helpers, having presumably tested the property market and found it hostile. One family of Minnesotan

(Overleaf) Dire Wolves in South America may have hunted cooperatively for Giant Sloths

Grey Wolves illustrates the complexities of dispersal. An adult female roamed over 4000 square kilometres (1500 square miles), alone, during 14 months of nomadism before she met a lone male with which she established a 69-square-kilometre (27-square-mile) territory. The following year the male was killed and his place (and three pups) were adopted by a lone male that had previously travelled a range of 300 square kilometres (120 square miles). Two of the three pups dispersed, but one female stayed to usurp her mother as the breeding female, a role she held for the next eight years. Some of that pair's offspring dispersed as far afield as 190 kilometres (120 miles), while others settled near by. Two females that settled in nearby territories abandoned their mates when their pups died, and returned to their parents' pack. One of these remained at home indefinitely, and the other moved on again after a few weeks. The decision to stay will be swayed by lack of vacant territories and also, probably, considerations of kinship. Even if the helper cannot breed, it will have a vested interest in the survival of its younger siblings.

Fathers, too, seek to maximise their surviving offspring, and among dogs, males contribute by tending pups diligently. This seems a marked contrast to most cat fathers. Two explanations commonly given for monogamy are that male assistance is necessary to rear the young or that males cannot monopolise more than one female. However, neither of these seems to be any more applicable to dogs than to most cats. A more obvious difference is that most canids eat a varied, omnivorous diet, including fruit, insects and carrion in addition to small vertebrates. A pair eating these foods will not necessarily interfere with each other and may even feed side by side, which may encourage cohabitation in a territory and a closer link between father and pups.

The lupine lineage prospered in the wake of climatic changes that began five to six million years ago. The replacement of forest by scrub and savanna accelerated. On the new plains there were new fleet-footed antelope, gazelle and zebra. Unlike their predecessors, the mega-herbivores, these fast and powerful creatures had soft bellies that could be torn by long-muzzled jaws. Just as these new prey stimulated the evolution of the pantherine cats, so they provided the first opportunity for fast, long-legged predators.

The first dogs to follow the ungulates on to the plains may not have been after their meat. Distinguished by its massive ears and flimsy teeth the Bat-eared Fox has come to depend on the insects that thrive on ungulate dung. With its astoundingly sensitive hearing, it can detect the underground raspings of a dung-beetle larva as it chews its unwholesome larder. The Bat-ear also feeds on harvester termites; when attacked, a column of these insects can dive for cover in less than two minutes and so the need to snap up termites at top speed has led to the evolution of a unique flange on the Bat-ear's lower jaw. The muscle rooted to that flange enables it to mince termites at the prodigious speed of more than three chews per second. It also has extra molar teeth to grind up insects. Since termites forage in huge numbers

Families of Bat-eared Foxes can forage together because their insect prey are replenished almost as fast as they are eaten

and vanish faster than they can be eaten, the Bat-eared Fox loses nothing by sharing the spoils. So, families of two or three adults with attendant cubs often forage together and when one hits the jackpot, its companions will run over to join in the brief, but succulent, feast.

Other dogs, those of the lupine line, grew and began to tackle larger prey. This new wave of running dogs hunted the new hoofed mammals of the plains. These dogs had speed and stamina, but were not heavyweights. The legs that bore them in swift pursuit of fleet quarry were unable to wrestle large prey, and their long muzzles were the wrong shape to stab or suffocate. The solution they arrived at was cooperation: hunting as a pack, they could kill large prey not by one bite, but by many. While the cats had relied on muscle power, the dogs' assault on the plains was based on pack power.

The ancestors of the pack-hunters probably already lived in family groups, just as many modern foxes and jackals do. So they would have been well accustomed to working together long before the arrival of large ungulates prompted them to embark on group hunting. Although their evolution was stimulated by the opening up of grassy plains, the pack dogs are principally hunters of the scrub where pack members can quickly out-manoeuvre a fleeing victim. Grey Wolves often hunt in

(Opposite) Black-backed Jackals in Namibia congregate at a carcass. Shareable food resources, like this, enable members of the dog family to capitalise on the benefits of group living

(Above) On Ellesmere Island, Grey Wolves co-operate in hunting Musk Oxen

sparse forests, as do Dholes, the red dogs of Kipling's *Jungle Book*, and the image of African Wild Dogs as hunters of open plains has been exaggerated because an open setting is the easiest in which to film them. Indeed, African Wild Dogs occur in dense mountain forests in the Bale Mountains of Ethiopia.

The lupine lineage was remarkably successful. From an ancestor named *Canis arnensis* came the Hare-eating Coyote, progenitor of the modern Coyote which, after flourishing in Europe, crossed back to North America two million years ago, along with various foxes and pantherine cats. *C. arnensis*'s line also gave rise, via the Etruscan Wolf, to the Grey Wolf, which was to become the most widespread wild mammal in the world, spanning Europe and Asia and crossing to America 700,000 years ago. The same stock produced the closely related Golden and Side-striped Jackals. An earlier lupine lineage produced the Black-backed Jackal and such pack hunters as the Dhole, which ranges from Siberia to south-east Asia, and the African Wild Dog whose ancestors had crossed to Africa five million years ago. This species' adaptations to long-distance running include the fusion of the middle two toes of the feet, a trait reputedly shared only with Israeli wolves.

The pack-hunters cooperate in the chase, their strength lying in numbers. Generally, the larger the prey to be tackled the larger the packs. Packs of Coyotes are more successful than pairs at catching hefty Mule Deer. Where the 70-kilogram (150-pound) White-tailed Deer is the main winter prey of Grey Wolves, an average pack contains about seven wolves (maximum 15), but where 350-kilogram (770-pound) Moose are the staple diet, packs average over nine members (maximum 23). Solitary wolves tend to kill smaller prey, such as young Caribou or smaller deer.

For pack-hunting dogs, surprise is of little importance: they start to run and, whether or not the prey sees them coming, they keep on running. The splendour of the hunting pack is the integration of individual prowess. Fast dogs curve in long encircling arcs, strong dogs fling themselves at flailing hooves, skilled dogs dash for the nose hold. African Wild Dogs may chase prey such as gazelle or Impala for 5 kilometres (3 miles) or more, at 60 kilometres (40 miles) per hour throughout. Occasionally they challenge larger prey, such as Wildebeest. Eventually, one may grab the Wildebeest's tail, and be dragged slithering and bumping through the dust until a companion grasps a skull-splintering hock. Then the skill comes into play as a specially experienced, agile and courageous dog flings itself at the head of the pounding Wildebeest ten times its weight and tries to grasp its upper lip or nose.

The highly endangered African Wild Dog is probably the most compulsively social of all canids

If it is successful, the Wildebeest stops in its tracks. Like their forebears, the modern dogs lack the precision bite of the small cats and the suffocating strength of the big cats. Instead, the pack's many jaws simply disembowel the prey, a gruesome but effective method.

The prey developed many tactics to ward off pursuit by the fleet-footed packs. For example, when Thomson's Gazelle see Wild Dogs advancing they fling themselves into pogo-stick leaps known as stotting, and easily misconstrued as panic. In fact, the gazelle are showing off. Their leaps must be exhausting, and all the more so the faster they are repeated. Stotting at a high rate requires great athleticism and so is an unfakable signal that the leaper is in prime condition. Since dogs have no wish to chase for miles after a gazelle that will eventually outpace them, the most impressive leapers among the gazelle are safe. Instead the pack will single out the slowest stotters and pursue them.

Wild Dog, Grey Wolf and Dhole packs are often larger than needed solely for cooperative hunting. African Wild Dogs can hunt alone if necessary. A single Grey Wolf is capable of killing even adult Moose, but no doubt finds it much safer, easier and more reliable to do the job with several companions. Nonetheless, of a pack of 15 Grey Wolves observed hunting Moose on Isle Royale usually only five or six actually made contact with the prey, although the others may have helped find and chivvy it. The most conclusive evidence that there is more to pack life than cooperative hunting is shown by the example of the Simien Jackal, which lives in packs but hunts alone.

The 16-kilogram (35-pound) Simien Jackals of Ethiopia are not closely related to other jackals, their old name, Abyssinian Wolves, being closer to the mark. They probably descended from a Wolf-cum-Coyote ancestor less than one million years ago, when their Afro-alpine habitat first emerged. The high meadows provided a paradise for rodents, supporting an incredible 40 kilograms per hectare (36 pounds per acre). It must have seemed a land of milk and honey to the ancestral Simien Jackals which, like all canids, would have been adept mousers. Unfortunately for them, subsequent climatic warming that started 1500 years ago caused the habitat to shrink to half a dozen mountain tops. These are now home to only 600 or so Simien Jackals, which live in packs defending territories averaging just over 6 square kilometres (2.3 square miles). Each pack consists of three to eight related adult males, one to three adult females (which might include immigrants), plus pups and juveniles.

Soon after dawn a pack of Simien Jackals will muster for a border patrol led by a large female. This female and the dominant male have particularly bright white tail-stocks that flash in the sunlight, and may be badges of status. At intervals the patrol will pause to cock their legs, and then nuzzle noses with tails wagging frantically. When the patrol is over, the Simien Jackals' springing gait settles to a meandering trot and the throng disperses. Soon lone Simien Jackals are hard at work, criss-crossing the meadows in search of grass rats. Occasionally one will sprint to intercept an unwary rat on its helter-skelter dash for the sanctuary of its burrow. If the rat escapes, the Jackal will dive at the burrow and dig, seeming to pivot all its weight on its hammer-shaped snout thrust deep into the entrance.

Relationships between neighbouring Simien Jackal packs are strained, and contests are decided by the numbers of animals on each side. While most young females disperse as two-year-olds, all young males stay with their natal group for life. The whole pack shares the burden of caring for the pups, regurgitating gut-loads of rodents for them until they are eight months old and skilled enough to hunt unaided. These helpers at the den may be useful, but pairs without helpers can nonetheless rear pups successfully. Strength of numbers in defence, their very restricted foraging habitat and the bleak prospect of dispersing animals seem sufficient incentives for Simien Jackals to form groups even though they never hunt together. Simien Jackal packs have much in common with large groups of Golden Jackals that live on large carcasses provided for them in an Israeli nature reserve. These Goldens stake their claim to the bonanza with piles of droppings ringing their territory like 'Trespassers will be prosecuted' signs. Like the Simien Jackal, they engage in daily get-togethers and patrols. Such social extravaganzas are characteristic of lupine dogs, at least those of Eurasian stock, but a rarity amongst the vulpine ones. For example, howling in chorus is at the heart of family life in all jackals, Coyotes and Grey Wolves. Although fox families keep in contact with barks and wails, and several individual voices may be heard replying to each other, they do not come together to call in chorus.

One cooperative trait that, above all others, does seem to unite the dog family is care of the young. Associated with this trait is family planning. Among group-living dogs, whether they be wolves, jackals or foxes, the general rule is that only one female breeds. The behaviour is so prevalent that it probably stretches back well before the split between wolf-like and fox-like dogs. The breeding matriarch is usually the oldest female in the group, and often the mother of her entourage of helpers. Where there is more than one male, the dominant probably fought his way to the top. How the rest of the pack are persuaded to give up the chance to breed is a puzzle that has fascinated researchers.

Each dominant male polices those of its own sex, preventing them getting ideas above their station. In a Grey Wolf pack the dominant male blocks sexual approaches by subordinates to the dominant female, just as she thwarts liaison between subordinate females and the dominant male. Both dominants interrupt flirtations in the lower ranks, and higher-ranking subordinates interrupt matings amongst those even lower in the pecking order. It was thought that this sexual policing might be made easier by a reluctance to inbreed, since most wolf packs are extended families.

The Simien Jackal of Ethiopia should be re-named the Simien Wolf to reflect its evolutionary origins

However, successful incestuous matings of every sort have now been observed in Grey Wolves, African Wild Dogs, Simien Jackals, Red Foxes and Bat-eared Foxes.

Amongst Simien Jackals the dominant male is unable to monopolise mating with his partner. Within the group, the dominant female will only accept sexual advances from the dominant male. However, she will readily mate with the males of a neighbouring group. Although each dominant male drives away neighbouring rivals in an endeavour to avoid being cuckolded, two thirds of matings seen so far have been by interloping males, including some 2-year-old subordinates.

Studies of a captive Grey Wolf pack, on the other hand, have shown that the females at least can be very successful in monopolising breeding. The dominant or alpha female produced pups whereas the subordinates, even when four or five years old, did not. Analyses of the sex hormones revealed that the majority of subordinate females could have conceived. They were cycling normally and ovulated, just like the dominant female. Furthermore they flirted with males as much as the dominant female would allow. The main barrier to their reproduction was psychological: their bodies were ready for reproductive action but they were constrained to behave like neuters. Indeed, when the alpha-female of one pack died during the breeding season, the most dominant of her daughters came into heat the same day.

The physical readiness of subordinate females may have other advantages for the pack too. As many dog owners will know, when bitches ovulate but do not conceive they may have a phantom pregnancy. Such a female displays all the outward signs of pregnancy: swollen body and teats, preparing a den and even coming into milk. Vets have traditionally viewed this as a misfiring of the hormonal systems. However, it may actually be a mechanism to prepare the subordinate for her role as nanny and wet nurse.

There is some flexibility in the family planning. Sometimes two females will breed and may even cohabit in the same den, nursing their cubs communally. This has been seen in Grey Wolves, Golden Jackals, Bat-eared Foxes, Red Foxes and Bengal Foxes. However, as observed in species from Wild Dogs to Dingoes, such dual births can result in the dominant mother killing the subordinate's cubs, whereupon the bereaved mother becomes a wet nurse for the dominant's cubs.

The more rigid the hierarchy in a group, the easier it is for the dominant to manipulate the breeding of her subordinates. For example, in an Oxford suburb where Red Foxes lived on average for five years or more, only one vixen generally bred in each group. A stone's throw away, in Oxford City, traffic accidents limited the average lifespan to 18 months, and most vixens had cubs. Perhaps the hierarchy based on age that operated in the suburbs had broken down amongst the youthful city vixens, most of which were contemporaries. A rigid hierarchy occurs in African Wild Dog packs. In the Serengeti, Wild Dogs whelp when the rains bring the migrating Wildebeest to the plains to calve and the dogs feast. For the remaining nine months of the year the pack is forced into foot-slogging nomadism, ranging over 2500 square kilometres (1000 square miles) in search of food. After the first nine weeks of the pups' lives, the family never sleeps in the same place on consecutive days. There is fragmentary evidence that the low numbers of Wild Dogs may be partly due to Lions which seek out and kill Wild Dog pups. So the peripatetic lifestyle of these dogs may also be an adaptation to throwing such predators off the trail. A single breeding female dominates a pack of up to 20 in which males outnumber females by at least two to one. Males are the main pup-carers, and uncles have been known to rear five-week-old orphans. Up to 19 relatively underdeveloped young are born in each litter and need tending for 14 months.

African Wild Dogs are at the extreme of two trends concerning pups. First, larger canids tend to have larger litters containing less developed pups – exceptions being the large Maned Wolf which averages two hefty pups and the small Arctic Fox which may have many cubs. Smaller species generally eat smaller, more scattered foods and thus live in smaller groups. So each has to work harder to gather extra food for the young and, being smaller, is less able to defend helpless young. This may have pushed them into having more advanced cubs and therefore smaller litters. In contrast, big species may have plenty of flesh to carry home, plenty of helpers to carry it, and the corporate strength to repel enemies. They can afford smaller pups, and thus more of them. Incidentally, lupine dogs regurgitate to their pups, but vulpine species have only been seen carrying food home.

The other trend is that the sex of helpers shifts from mainly females in small species, through equality in medium-sized species to predominantly male in some of the largest, culminating in the African Wild Dog. Those with predominantly female helpers are not only small, but also all belong to the vulpine lineage. Perhaps this trend can also be traced to the different foods of small and large canids. Because of their small prey, small foxes forage alone and do not rely on mob tactics to hunt and defend prey. They live in small, loose-knit groups. A subordinate vixen that stays at home could rear a litter unaided, and occasionally does so where food is plentiful, such as amongst the sea-bird colonies of Alaska's Aleutian Islands. Fox cubs are self-sufficient in three months and their decentralised social lives may make it easier for a subordinate female to evade the dominant's control. In contrast, because Wild Dogs hunt large prey that can only be killed by teamwork, they depend on each other for survival and live in tight packs. The dominant female's huge litter is tended by the entire pack for 14 months. She brooks no rivals and a subordinate would be unable to rear cubs unaided. Subordinate females therefore may have less to gain by staying at home in Wild Dog than in Red Fox society.

At least sometimes it may not be a disadvantage to a dominant female fox for her subordinate to breed. In contrast, a dominant female Wild Dog needs the entire energy of her pack to feed her pups. This may lead her to exert disproportionate pressure to evict subordinate females. Furthermore, the Wild Dog's lifestyle is compromised by motherhood, each additional mother both weakening the pack's workforce and increasing its burden. In contrast, the Red Fox's hunting is less affected by motherhood, which is also compatible with its cooperative activities such as communal nursing, adoption, den guarding and keeping the cubs warm. Grey Wolves are even bigger than Wild Dogs, but do not have such a bias to male helpers, perhaps because they are less dependent on pack-hunting.

The sexes will fare differently if they disperse, both as aspiring territory-holders and as wayfarers. Wayfaring males may secure matings as wanderers, but a pregnant female wanderer is unlikely to prosper. Thus, if all else is equal at home, males may be more likely to set forth than females. However, all else is not equal for Wild Dogs: cuckoldry is ruled out because oestrous females are surrounded by their pack of protective males. Within his pack, however, a junior male Wild Dog may occasionally sneak a mating when the dominant is distracted. Once a society has started down the road to a biased sex ratio this tendency becomes self-sustaining: an adolescent male Red Fox setting forth will find a vacancy where one resident male has died, but for an adolescent vixen to find a vacancy might require all of three or four females to have died in one territory. The reverse applies to Wild Dogs. The fact that it takes several males to rear one female Wild Dog's litter makes it advantageous for parents to produce teams of sons, which may explain why even at birth 59 per cent of their pups are male.

The African Wild Dog's closest living relative, albeit very distant, is the Bush Dog. The ancestors of this and other South American canids began trekking to their present home two million years ago, when the Panamanian isthmus opened. One lineage that travelled south gave rise to eight species from the Crab-eating Fox in the north to Darwin's Fox on the island of Chiloe in the south. These eight species have been called foxes since the days of the conquistadors, but recent molecular studies have proved that they actually descend from lupine ancestors. Their behaviour reflects this origin. For example, Crab-eating Foxes growl in a most un-vulpine way and the young of both sexes may remain in their parents' territory for at least two years. This is common in the wolf-like side of the family but in the fox branch only daughters generally postpone dispersal for so long. Another lineage which had split from that of these 'foxes' six million years ago produced the ancestor of the Maned Wolf, which has the lupine habit of regurgitating to its young. A third lupine strand produced the ancestor of today's Bush Dog.

The 23-kilogram (50-pound) Maned Wolf is a denizen of open plains where its stilt-like legs carry it over long grasses. It feeds on rodents, supplemented by fruit, and thus lives in pairs. In contrast, the 5-kilogram (11-pound) Bush Dog lives in packs, is an inhabitant of dense Amazonian jungle, and stands only 25 centimetres (10 inches) at the shoulder. With their small size, rounded ears, short legs and

The Maned Wolf's long legs may adapt it to hunting rodents in tall grass. In Brazil it also eats a fruit, Solanum lycocarpum *(which translates as Wolf Fruit). Apart from its nutritional value, one suggestion is that Wolf Fruit acts as a natural 'wormer', purging the Maned Wolf's gut parasites*

stumpy tail they would easily dodge under low vegetation in their jungle home. These muscular, heavy-jowled dogs are the closest in shape of any wild dog to an otter, and have webbed feet and can swim well. They whistle like stricken guinea-pigs, probably to keep in contact in the dense undergrowth.

To judge by observations of a captive pack, Bush Dogs are compulsively sociable. The seven members of the captive pack spent the night in a sleep heap, their long-low bodies wound in a loose knot from which a miscellany of heads, legs and tails protruded. Their day began with a ceremonial whistling before they set off to patrol their territory in single file, with an old male in the lead. The dominant female initiated marking at a log arch, swinging her hindquarters upwards to spray urine on the roof of the arch and soaking her tail in the process. The other pack members marked too, each male arching its back, pointing one hindleg forwards and bending awkwardly to spout urine forwards and upwards to the log arch. As the pack milled back and forth each was showered by the cocktail of urine dripping from the arch. So the animal, as well as their territory, were imbued with the pack scent.

When a dead calf was placed in their paddock the whole pack carved meat from the carcass with slicing teeth. Each dog used the others' bodies as a lever, gnawing centimetres away from a companion's jaws with neither a growl nor a snap. Nobody has observed Bush Dogs in the wild in detail. However, their astounding harmony in captivity, and the way they bunch in a throng at the sight of prey, suggests they at least sometimes hunt as a pack.

An intriguing feature of the Bush Dog is the very sharp rear blade on its scissor teeth. In most dogs from Red Foxes to Grey Wolves the scissor tooth has a meat-slicing blade at the front and a semi-circular basin to the rear for grinding. However, three widely spaced species have abandoned the basin in favour of a second, rearward, cutting blade – the trenchant heel – and have also lost much of the grinding capacity of their back teeth. The three are the African Wild Dog, the South American Bush Dog and the Asiatic Dhole. They also share the unusual feature of whistling or twittering rather than howling, which at least for the latter two may help them coordinate while encircling prey in dense cover. Molecular studies reveal that the slicing three, along with Black-backed Jackals, are distant relatives sharing descent from an early lupine branch. Perhaps their shared ancestor had trenchant heels or perhaps the three dedicated meat-eaters, having little use for a grinding basin, invented the extra blades independently. A flesh-cutting heel is a cat-like trait, and it evolved as the dog's solution to the problem of fleet prey that cats solved with the evolution of the pantherines. Today, the only dogs with sharp-heeled teeth are the predatory pack hunters, so it seems a fair guess that any extinct sharp-heeled dogs were also sociable.

The lineages that produced the Bush Dog, Maned Wolf and South American 'foxes' probably all arose from early lupines that had never left America. A fourth group of southern travellers came from Eurasian wolves that returned to their ancestral home of North America about two million years ago. The most widespread of these, the Dire Wolf, thrived for much of the Ice Age. Starting about 1.6 million years ago, the Ice Age has so far comprised up to 17 cold-warm cycles, each spanning about 100,000 years. At the end of the last cool period, 10,000 years ago, the large herbivores on which the Dire Wolf preyed became extinct, taking their predators with them. Another Wolf, the Falkland Island Wolf, was closely related to Coyotes and probably drifted to the Falklands on ice floes from Cape Horn while feeding on seals and sea-birds. It became extinct in 1876.

The Ice Age affected the fox side of the family too, as illustrated by the history of the Swift Fox. The ancestors of this 2–3-kilogram (4.4–6.6-pound) fox crossed back to North America little more than half a million years ago and adapted to arid lands. Unusually among canids, the Swift Fox hunts alone for prey larger than itself, pursuing jackrabbits at speeds that can top 50 kilometres (30 miles) per hour. During cool periods of the Ice Age rolling glaciers pushed south, converting the fox's prairie home to tundra. In response, 250,000 years ago the Swift Fox gave rise to the Arctic Fox. Retreating from the ice to warmer climes the Swift Foxes survived in enclaves which began to evolve into separate species: the Colorado and Nebraska Swifts. Often called kit foxes, these species are as different from each other as either is from the Arctic Fox, and the Alberta Swift may also constitute a separate species. Modern kit and swift foxes are smaller than their ancestors of 11,000 years ago – presumably, like Ruppell's Fox, so that they can survive in areas too arid for the competing Red Fox. The balance between Arctic and Red Foxes is so fine that even small changes in climate, such as the warming of the northern hemisphere over the last century, can have a noticeable effect. For example, between 1880 and 1940 the mean annual temperature increased by 2.5° C (4.5° F) in Spitzbergen, bringing it a climate previously enjoyed 1500 kilometres (900 miles) to the south. This led to an increase in prey species, allowing Red Foxes to invade territory previously the refuge of Arctic Foxes. The invading Red Foxes pushed Arctic Foxes out of the warmer, low-lying coastal areas where the blue form predominated to high-altitude inland sanctuaries where the white form has an advantage. In consequence, the balance of Arctic Fox colours has changed in Scandinavia: where previously blue individuals were common, they now number less than 1 per cent of the population.

Having survived the Ice Age, modern swift foxes faced a more serious threat: humans. European settlers arrived on the Canadian prairies in 1627 and by 1938 the last of the Alberta Swifts had been shot in Canada. Attempts to reintroduce them had resulted in 15 breeding pairs by 1991, split between two sites in Alberta. However, in the intervening years the Red Fox had spread and it remains to be seen whether the Swifts can compete with this large newcomer to the prairies.

(Overleaf) The Bush Dog's low-slung body, short ears and stubby tail are probably adaptations for hunting in dense forest. Pack members twitter to keep in contact, and rest together in a sociable sleep-heap

The pack hunters of the plains suffered at least as much at the hands of humans. As well as destroying the herds they preyed upon and the habitat that supported them, people viewed all but the tiniest canids as competitors. With a terrible irony we took the dog into our homes and cosseted it, while relentlessly striving to exterminate its wild kin. In Europe, wolves and other large canids have been virtually annihilated. Just over a million years ago African Wild Dogs occurred in Europe as well as in Africa. A hundred years ago their range had shrunk but they still inhabited almost every part of sub-Saharan Africa except tropical rainforest. Now they number under 4000 on the whole continent. In 1988 four marriageable sisters dispersed together from a Wild Dog pack in the Serengeti, and wandered for a year without finding an eligible male, because there is none left. We can weep for them not only as the remnants of a splendid species, but also as individuals as full of character as our cherished pet dogs.

The effect of human pressure on some canids has been rather like a journey back through time. A Coyote-like dog in Eurasia gave rise to the ancestors of the Grey Wolf, which returned to North America. At the peak of the Grey Wolf's success it drove its humble Coyote cousins to backwaters, such as the deserts of Mexico, or at least to the periphery of its territories. Then people changed the balance. The cutting-down of forests, the spread of agriculture since 1800 and hunters together virtually eliminated the 50-kilogram (110-pound) Grey Wolf. In the 30 years after 1865, hunters killed nearly every wolf from Texas to the Dakotas, from Missouri to Colorado. In the hunters' wake came the 8–20 kilogram (17.5–45-pound) Coyote, invading its larger relative's haunts. The Coyote's smaller size allowed it both to hide from people and to live on the mice and squirrels that survive modern farming.

Now the Coyote is invading the Grey Wolf's range, and the two species remain sufficiently similar to inter-breed. Where their populations are fragmented, lonely male wolves may easily intimidate puny male Coyotes and find solace in the company of the numerous female Coyotes that are invading their range. Their progeny are increasing in a 500-kilometre (300-mile) hybrid zone of 'Coywolves' around the Great Lakes. The Coywolves might integrate into either wolf or Coyote population, but in fact only wolves with Coyote blood are found, not Coyotes with wolf blood. Geneticists can determine which species was mother or father of the hybrid, and it seems that the process is one-way: male Coyotes do not form successful liaisons with female wolves. The coyotisation of America's wolves follows the spread of agriculture, and thus of Coyotes, beginning in Minnesota and spreading to Quebec and Ontario in the East.

Members of the dog family look and live alike, and their similarity in appearance is more than skin deep. For example, Coyotes in the southern states have cross-bred with the rare Red Wolf for decades. This cinnamon-coloured wolf formerly

The Coyote is better suited to living alongside people than is the Grey Wolf, so in the last 200 years the Coyote's range has expanded where the wolf has been exterminated

ranged from southern Florida to central Texas and perhaps beyond. Like the Grey Wolf, it suffered from predator control programmes and twentieth-century agriculture. The Red Wolf also mated with Coyotes but, in this case, cross-bred itself into oblivion. By 1970 seemingly pure Red Wolves were confined to south-western Louisiana and the south-eastern corner of Texas, and by 1980 they were extinct in the wild. Fortunately in 1977 a breeding programme had begun and some 79 candidates had been captured and evaluated for it. Most were considered to be Grey Wolves, Coyotes or 'Coywolves' and only 14 were judged genetically pure Red Wolves. By 1988 these 14 had produced 80 Red Wolves in captivity at eight locations in the US, and a group of them were reintroduced to Alligator River National Wildlife Refuge in North Carolina. In November 1991 they were reintroduced to the Smoky Mountains, and a pair has been released on each of three islands off Carolina, Florida and Mississippi

While hailed as a conservation triumph at the time, this rescue has since become the centre of a philosophical debate. Scientists analysed the genetic make-up of Red Wolves that had lived between 1905–1930 using skins that had lain fusty in museum vaults. They also analysed all the candidates for the captive breeding programme, from stored blood samples, and the descendants of the 14 judged to be Red Wolves. Their studies showed that none of the candidates rejected as Grey Wolves was in fact a Grey Wolf, being mainly Coyotes or Coywolves. Furthermore the Red Wolves – those that had lived from 1902–30 and the modern captive-bred ones – had no genetic characters that could not be found in either Grey Wolves or Coyotes. In fact, they were genetically indistinguishable from Louisiana Coyotes. So, either the Red Wolf had already cross-bred itself out of existence by the late nineteenth century, or it never existed as a true species but was always a hybrid between the Grey Wolf and Coyote. Whichever explanation is true, one of the biggest ever conservation projects may have struggled to save a species which already did not exist. However, on the principle that a rose by any other name would smell as sweet, the important thing is that there is once more a big red canid roaming wild in the southern states.

Human pressure on the canids is increasing and now even the Coyote is too big for the modern world. The current chapter in the dog family's story is a replay of the first, with small, nimble opportunists flourishing. Today the pinnacle of canid success is the 6-kilogram (13-pound) Red Fox. This canid is the most widespread of all the wild Carnivores, adapting to desert, tundra, farmland and city, and prospering on foods from beetles and berries to beefburgers. The same build, and doubtless the same opportunism, that enabled the pioneering early dogs to prosper alongside fearsome cat-like sabre-tooths now enables the Red Fox to dodge the traffic in the busiest cities in Europe.

Family Tree: The Dogs

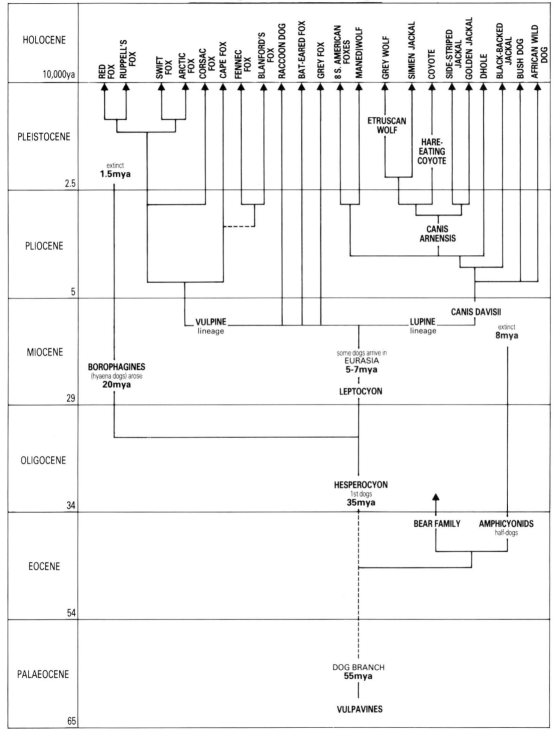

CHAPTER 4

RICH MAN'S TABLE

Hyaenas, or at least the Spotted Hyaenas which characterise the family in most people's minds, have a terrible reputation. According to African mythology, witches ride their backs. Colonial mythology deemed them loathsome scavengers. They are pictured fretting, excluded from the belly-bloating gore of a Lion kill, lurching malevolently at jackals. Why we should so despise scavenging is not clear, since it is a lifestyle at which our own ancestors were adept. In addition, like many popular images, that of hyaenas is distorted.

It is not, and never has been, true that the Hyaenidae is a family distinguished by scavenging. Of the four species alive today – the Aardwolf and the Spotted, Striped and Brown Hyaenas – only the latter two are full-time scavengers. Futhermore, members of all eight Carnivore families will scavenge given the chance, and many make a living from it. Specialist insect-eaters, such as Banded Mongooses, will nibble at a buffalo carcass. Even the bamboo-eating Giant Panda will gnaw on a carcass now and again. Wolverines in some regions live almost solely on carrion, and Black Bears compete with them for pickings. Both Raccoons and Red Foxes can thrive on mankind's left-overs. However, such run-of-the-mill scavengers lack the ability to crack bones, shared by most hyaenas. The bone-cracking teeth of the Spotted Hyaena can exert a pressure of 800 kilograms per square centimetre (11,400 pounds per square inch), equivalent to that of an elephant wearing stiletto heels. Behind these are massive carnassials that can shear through hide and tendon, and one can almost hear the gurgling in the cauldron of the acidic stomach as it converts bone to powdery calcium.

Although the slightly smaller Striped and Brown Hyaenas are similarly equipped, bone-cracking does not unite the hyaena family. The fourth species, the Aardwolf, has more dainty feeding habits and could scarcely snap the flimsiest bone. In fact, all the four species have in common are detailed features of their anatomy, particularly the bones of their inner ears, and special scent-secreting pouches beneath the tail which produce perfumed paste said in Africa to be witches' butter. The pouch is least elaborate in the Spotted Hyaena. The four species turn the pouch inside out during social encounters, and regularly daub the perfumed paste throughout their territories.

The bone-cracking prowess of the Striped, Brown and Spotted Hyaenas might lead one to assume that the ancestral hyaenas had similar abilities, and that the Aardwolf has strayed furthest from its roots. If anything, the reverse is true. The hyaena's story began in the jungles of Eurasia about 22 million years ago, when the hyaenas' ancestors and other early representatives of the cat branch still spent much of their time in the trees. The first ancestral hyaenas probably looked, and lived, much like the modern Banded Civet, travelling their territories alone by night, snapping up frogs and lizards, raiding birds' nests, gleaning insects and occasionally eating fruit. They would also have taken advantage of the many foods that dropped to the floor below, including fallen carrion. Eating foetid food requires a strong stomach which was an important prerequisite for the scavenging of later hyaenas. Another talent of the hyaenas, which also seems widespread in cat-branch Carnivores, is the ability to regurgitate on demand, and this may have evolved as a way of getting rid of toxic or sharp debris.

Brown Hyaenas in the Kalahari desert travel prodigious distances in search of carrion. Where the carcasses of large antelopes abound more Browns can share the spoils and consequently they form larger clans

The mainstream of hyaena evolution involved species like Ictitherium *which were more reminiscent of dogs than of modern hyaenas*

One of the earliest hyaenas to be exhumed to date is *Plioviverrops*, a lithe, civet-like creature that lived in Eurasia between 22 and 20 million years ago. It is identifiable as a hyaena from details of its middle ear and dental structure that are shared by all members of the family. However, in life *Plioviverrops* was probably most similar to a modern African Civet. Like this animal, *Plioviverrops* may have moved languidly down sloping boughs, grasping with cat-like claws, and then leaped silently to the leaf litter below to feed on small vertebrates, ripe fruit and such gastric challenges as foetid carrion, toxic millipedes and poisonous plants. Its lineage prospered, and produced descendants with more pointed jowls and racy legs, as the dog family had done in North America. These descendants became dog-like wandering generalists.

As dogs, hyaenas were a great success. In their heyday, 15 million years ago, 30 or more species of dog-like hyaenas flourished in the Old World. They were not specialist bone-crushers, but nimble-bodied, wolfish creatures, fashioned from the raw material of the cat branch. Amongst them was *Ictitherium viverrinum*, which was moulded into the shape of a jackal. Some of its relatives were more heavily built and wolf-like, others were more like Coyotes, and others more like foxes. At some Miocene sites, fossil remains of *Ictitherium* and similar species outnumber those of all other Carnivores put together.

The dog-like hyaenas had dog-like molar teeth that would have allowed them to supplement meat with fruit and insects. Perhaps, as with dogs, this diet allowed territorial pairs to cohabit. Certainly, there was no difference in size between the sexes, as is associated with the exclusive territorial polygyny of cats (see page 45). Some dog-like hyaenas probably lived in groups like wolves and African Wild Dogs, and the parallel with true dogs suggests that in their groups breeding was probably the prerogative of the dominant pair.

The mainstream of hyaena evolution in Eurasia and Africa was paralled by the story of the dogs in North America. Then, about five to seven million years ago, the true dogs migrated out of North America, travelling across the Bering landbridge to Eurasia, and the Old World dog-like hyaenas crashed. The close coincidence between these two events makes it tempting to conclude that the dogs, full of competitive vigour from North America, out-shone their Old World equivalents. However, by the time the true dogs were diversifying and spreading through the Old World, five million years ago, the dog-like hyaenas were already well on their way to oblivion. Their decline seems to have been caused by changes in world climate that occurred at about the same time as the dogs were invading Eurasia. Once the climate had stabilised, the dog-like hyaenas might have bounced back, but by then the dogs had taken their places. Today there are some 22 species of canids in the Old World occupying niches that were once held by dog-like hyaenas.

The climatic changes that toppled the dog-like hyaenas had the biggest impact in the Old World, but even in North America the diversity of true dogs fell by more than a third, and the hyaena-dogs disappeared. In addition, many herbivores and older members of the cat family became extinct. No one knows why the changes in climate were so catastrophic, nor even exactly what they were. However, 13 per cent of mammal families and 73 per cent of genera became extinct over a period of about 1 million years. Fossil leaves indicate that the northern hemisphere warmed from a trough seven to eight million years ago to a peak five to six million years ago, and then cooled again. This coincided with the Messinian Crisis when the Mediterranean Sea dried up and the door to Africa was opened. However, the changes in mean annual temperature would have had different effects at different latitudes. The mosaic of seasonality, rainfall and temperature and their effects on habitats in different parts of the late Miocene world are still too complicated to untangle. What is clear is that in the aftermath of the turbulence, forests were replaced by scrub and open savannas, and these prompted the evolution of new fleet-footed, plains-dwelling ungulates. Fertilised by the ungulates' dung, the plains sustained copious insects, amongst them the termites. This was greatly to the advantage of the only one of the original dog-like hyaenas which has persisted in its old trade. The Aardwolf survived because 15 million years ago its ancestors

specialised on termite-eating and it has successfully resisted the only canid challenger to this trade, the Bat-eared Fox.

The Aardwolf is a jackal-sized, black-and-fawn striped creature, found in southern and east Africa but missing from a 1500-kilometre (900-mile) wide band in between. Despite some shaky links, its lineage can be traced directly to *Plioviverrops* at the very roots of the hyaena family. Today, Aardwolves are dedicated eaters of just two genera of termite. Their mainstay is a harvester termite, *Trinervitermes*, dense columns of which emerge at night to harvest dry grass and carry it back to their dome-shaped mounds for later consumption. Unfortunately for the Aardwolf, *Trinervitermes* cannot stand the cold at night or sun by day, and so does not venture out in winter. The Aardwolf is saved from starvation by turning to a hardier termite, *Hodotermes*, which does emerge in cold weather by day, being dark-pigmented to protect it against the sunlight. Nonetheless, during June, the southern mid-winter, adult Aardwolves lose a quarter of their weight and juveniles, then about six months old, sometimes starve. The lack of grass-harvesting, surface foraging termites in the wooded habitats of the wide band between east and south Africa is responsible for the absence of Aardwolves in that zone.

A foraging party of *Trinervitermes* might number 4000, a sizeable enough snack to tempt many predators. However, about one-third of the party will be soldiers, which flank the columns of workers busily heaving grass back home. The soldiers' noses are loaded with chemicals called terpenes, and they will squirt these in a sticky goo through their snouts at any adversary. This defence is more than sufficient to deter other termite-eaters, such as Bat-eared Foxes, White-tailed Mongooses, Pangolins and Aardvarks. Aardwolves seem resilient to the stomach cramps that terpenes normally induce. The strong guts that allowed their ancestors to cope with foetid carrion were probably a useful step towards tolerating noxious termites. However, the Aardwolf does have to detoxify the terpenes and this complicates the process of digestion. In addition, termites come mixed in with sand and prickles, which further diminishes the food value of each meal. This may explain why the Aardwolf's metabolism runs at a slow rate relative to other mammals, including other hyaenas. A thrifty expenditure on bodily energy slows down all bodily growth, and that includes the growth of babies. Thus its diet constrains the Aardwolf to breeding slowly: after a gestation of 90 days its litter generally numbers two or three. However, the young cubs grow relatively fast, perhaps in order to achieve adult size before the rigours of winter food shortages.

The Aardwolf is equipped with a grotesquely long, broad tongue and huge salivary glands that wash a column of termites into a sandy gruel. On a good night

The stripes of the Aardwolf probably reflect the ancestral patterning of early hyaenas, although one unlikely suggestion is that they are a recent device to mimic the larger Striped Hyaena

Aardwolves are highly specialised to a diet of just one type of termite

a single Aardwolf can lick up 300,000 *Trinervitermes*, giving a startling total of 12 million a month. Since overwhelming termites requires nothing more than a slobbery tongue, the Aardwolf's molar and premolar teeth have degenerated to flimsy pegs. However, it still retains fearsome canines, because individuals still need weapons to fight for territories and mates.

Trinervitermes pop out of their burrows unpredictably, so it takes each Aardwolf about six hours to find its nightly quota. To ensure a good chance of securing sufficient food, a pair of Aardwolves defends a 150–200-hectare (370–500-acre) territory encompassing about 3000 *Trinervitermes* mounds. Since there is little scope for collaboration in overpowering a termite, the members of a pair hunt alone. Their territorial system is stable for most of the year but partly breaks down during the two-month mating season. At the beginning of this, the toughest males start exploring the neighbourhood, avoiding the territories of other dominant males but trespassing freely on those of weaker subordinates. Far from being surreptitious, the dominants paste scent marks everywhere they go. After a month of escalating chaos, the female Aardwolves come on heat, and their scent-marking of territory

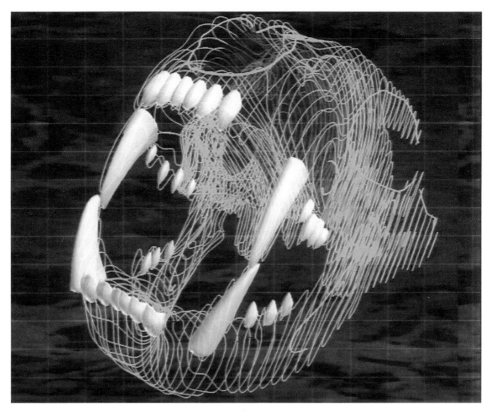

Aardwolves retain impressive canine teeth for social reasons, but their cheek teeth are reduced to irrelevance

borders increases to an almost frenzied rate. Females cohabiting with low-ranking males further flaunt their perfumes during sorties into the territories of dominants. Having mated first with his own female at home, a dominant male sets off in search of neighbouring females, guided by their scent marks. When he finds one he drives off her companion, sometimes interrupting him in the act of mating. Then the dominant prances around her with tail erect, displaying his pouting anal gland.

Aardwolf copulation is a protracted affair, lasting up to four and a half hours. The first ejaculation occurs about one or two hours after mating begins, and subsequent ejaculations follow at hourly intervals. Indeed, it seems that the only way to draw the proceedings to a close is for the female to enter her burrow, thereby brushing the male off her back. In apparent anticipation of this unceremonious interruption, dominant males carefully lead females away from their den before mounting them. The prolonged mating may help the male ensure that his sperm have a good start over those of any other male that is likely to be waiting to mate with the same female.

The Aardwolves' system allows a female cohabiting with a subordinate to mate with the best-quality male around and so give birth to sons that might be similarly prolific and produce many grandoffspring. It also carries the risk that the cuckolded subordinate might desert, which would be disastrous for her and her cubs. Both male and female are needed to rear the cubs, one parent defending the den from marauding jackals while the other forages. Indeed, a diligent father may baby-sit for six hours a night. There is a case on record of a male deserting an unfaithful mate. This occurred when several males had been accidentally killed during a campaign to poison locusts, leaving widowed females. A surviving local dominant mated with most of these and with the partner of a weaker male. The cuckolded subordinate male left and moved in with the only solitary female that had not been mated by the dominant. However, usually such a female is not available and subordinate males have no choice but to tolerate cuckoldry. Leaving home would rob them of the companion and territory needed to survive the remaining months of the year. In addition, an unfaithful female will also mate with the subordinate male, introducing uncertainty over paternity, and this may be a tactic to forestall defection. If so, it appears to be only partly successful, because males with unfaithful partners are generally less diligent babysitters than dominant males.

The Aardwolf's system shares features of that of small cats, in which males roam the territories of several females but ignore their kittens, and that of most dogs, in which monogamous males share their mate's territory and are diligent fathers. The Aardwolf's compromise seems to be beneficial to all concerned: dominant males get most matings, more cubs get the best fathers, mothers get help with rearing and subordinate males have some hope of mating and a territorial stake. This system probably arises because the erratic nature of the termites' emergence, and the minimal impact of Aardwolves on their numbers from night to night, probably means a male can share a female's territory with little or no expansion.

The other modern hyaenas have different social lives from the Aardwolf and follow a very different trade. Their bone-cracking teeth, lacking in the Aardwolf and its dog-like hyaena ancestors, enable them to be master scavengers. Among the mammals, the first master scavengers were probably some of the creodonts (see page 23), such as *Sarkastodon* of eastern and northern Eurasia. They were superseded by a mysterious group called the percrocutoids, which probably originated from the dog branch. The creodonts survived longest in western Eurasia and Africa, perhaps because it took the percrocutoids longer to spread there from the dog-branch's North American base.

The percrocutoids lived alongside the early dog-like hyaenas and relations between the two were probably rather strained, just as they are between bone-cracking hyaenas and jackals today. One can imagine a gathering of *Ictitherium*, nibbling daintily at sinew on a carcass, being usurped by the arrival of a large, heavy-headed *Percrocuta*. This percrocutoid could crack open marrow bones with massive jaw power. Other *Percrocuta* might also have arrived at the carcass from down-wind, their phenomenal sense of smell having guided them to the feast.

Percrocuta *represents a mysterious group of bone-crunching Carnivores that were eventually replaced by hyaenas*

Perhaps they would have been well acquainted, members of a loose-knit society accustomed to such gatherings of the clan.

Until about 15 million years ago, the percrocutoids and the hyaenas both prospered, but then the percrocutoids began a 10-million-year decline. The start of this coincided with the development among some hyaenas of bone-cracking teeth. The innovative new hyaenas were in a tiny minority compared to their dog-like cousins, but they competed directly with percrocutoids. The smallest, jackal-sized percrocutoids were the first to go, the bone-cracking giants holding out until about seven million years ago. The coincidence of the hyaenas' rise and the percrocutoids' fall is so precise that it seems likely that they are cause and effect.

By twelve to ten million years ago there were two distinct strands of hyaenas in Eurasia: the mainstream dog-like hyaenas and the avant-garde of bone-crackers. Even in those early days, separate ancestors were recognisable amongst the new bone crunchers that would one day lead to the Brown, Striped and Spotted Hyaenas of today. Doubtless, like dogs, the dog-like hyaenas did a lot of scavenging, and probably, like modern Spotted Hyaenas, some bone-crackers were also adept hunters. Carrion was probably crucial to them both, and both were mobile opportunists. Then, the change in world climate of seven to five million years ago (see page 121) devastated the dog-like hyaenas. The percrocutoids too finally tumbled to extinction, but a handful of species of bone-cracking hyaenas survived to take their places.

While the bone-cracking hyaenas were rising to ascendancy in Eurasia, master scavengers of a very different stock were flourishing in North America. In a remarkable parallelism, the Old-World hyaenas produced dog-like forms and the true dogs of the New World produced hyaena-like forms. These borophagines or hyaena-dogs became even more committed to bone-cracking than their hyaena counterparts (see page 118). Their establishment in the New World probably prevented the bone-cracking hyaenas from invading North America via the Bering land bridge. The hyaena-dogs lasted until about 1.5 million years ago, when wolves, which had returned from Eurasia, could splinter bones of the sizes available in North America.

While the true dogs were invading Eurasia, one dog-like hyaena did cross the other way. Its name, *Chasmoporthetes*, translates rather quaintly as he who saw the canyon. In North America, wandering omnivore and bone-cracking niches were monopolised by the true dogs, but *Chasmoporthetes*' descendants found an opening and evolved to become the first cheetah-like sprinting hyaenas. Although Cheetah-like cats reached North America three million years ago, a really fast one did not evolve there until 20,000 years ago. *Chasmoporthetes* died out about 1.5 million years ago.

In the Old World, the bone-cracking hyaenas had emerged as the dominant scavengers by five million years ago. The mega-herbivores were huge vegetarians, unassailable to all but the mightiest sabre-toothed cats. But when these giants died, their corpses provided rich pickings for a new generation of hyaenas. One of them, *Pachycrocuta*, was a 200-kilogram (440-pound) mega-scavenger whose sledge-hammer teeth could splinter the marrow bones of an elephant.

Pachycrocuta (above right) was a heavyweight predecessor to Spotted Hyaenas

Pachycrocuta's teeth (below right) show how hyaenas combine huge carnassial flesh-cutters and sledgehammer bone-crushers

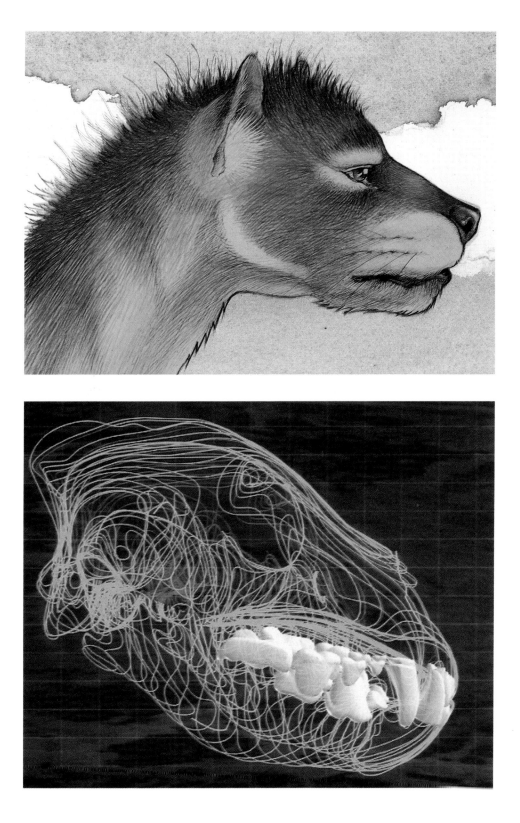

Perhaps Pachycrocuta *disputed kills with the prehistoric cat,* Homotherium, *just as Spotted Hyaenas and Lions are rivals today*

Pachycrocuta's teeth illustrate the basis of modern hyaenas' success. Its upper and lower third premolar teeth were huge conical crackers, and a third bone-holding cone jutted up from the fourth lower premolar tooth. Combined with massive jaw muscles and special vaulting to protect the skull against the huge forces, these teeth could grasp and splinter marrow bones. Crucially, *Pachycrocuta* also had magnificent carnassial teeth for slicing meat, allowing it to demolish 25 kilograms (55 pounds) of flesh and bone a day. Its whole jaw was shortened to increase the power of the closing hinge. Consequently, as in the short-jawed cats, the rearward molars found in dogs and viverrids have gone from these hyaenas. By organising their teeth so that the premolar bone-crackers do not interfere with the meat-slicers to the rear, hyaenas can smash a bone with the sledgehammer teeth without blunting the blades of its scissors. Percrocutoids also used their premolars for bone-cracking, but hyaena-dogs used their scissor teeth (see page 82).

Pachycrocuta depended on large herbivores, and with their demise in the late Ice Age, its place was taken by a scaled-down version, the Spotted Hyaena. During warmer weather between glaciations 35,000 to 15,000 years ago, Spotted Hyaenas were depicted in cave paintings in Europe. Despite being lightweights in comparison with *Pachycrocuta*, they can demolish carcasses with devastating speed. They bolt flesh, bone and sinew into large, fiercely acidic stomachs, the record being held by a mob of 35 that took just 36 minutes to consume a whole zebra. However, their trade is a dangerous one. Of Striped and Spotted Hyaena skulls in museums, 35 and 40 per cent respectively have broken teeth, the next highest figure being 29 per cent for Grey Wolves, which also do a lot of bone-cracking. Among Cheetah, which slice meat off the bone, only 17 per cent had broken teeth. Furthermore, three-year-old Spotted Hyaenas have teeth as worn as those of six-year-old Lions, despite the fact that hyaenas have disproportionately large teeth to counteract wear.

Today there are three bone-cracking hyaenas, the Spotted, Striped and Brown, and there have been only three to six species of them at any one time since their take-over from the percrocutoids. This lack of diversity may be due to the lack of scope for differences in the bone-cracking trade: any bone a small hyaena can crack, a big hyaena can crack better. Indeed, a bone-cracking specialist needs to be as big as the prey bones of the day permit, since the skeletons of small prey can be eaten by carnivores without sledgehammer teeth. Size is also important in the intense rivalry between scavengers. Spotted Hyaenas will kill Brown Hyaenas when they meet at a carcass, and Lions kill Spotted Hyaenas given the chance. Probably because of this, there has only ever been one species of full-time scavenging hyaena in any one place since the late Miocene. The most similar and most dedicated to scavenging of the modern hyaenas are the Striped and Brown, and the two never coexist.

The Brown Hyaena of southern Africa is now the rarest of hyaenas. It travels alone in vast territories of 300 square kilometres (120 square miles) or more, where food is so scattered that it may roam 50 kilometres (30 miles) each night, and keeps in touch with its group-mates by leaving scent markings. The Brown Hyaena's anal pouch, which lies just below the base of the tail, produces a black and white paste. The hyaena can turn the pouch inside out. The pouch is partly divided into two lobes by a groove and coated in a white fatty secretion of sebaceous oil glands. On either side of the everted 'bubble', other glands pour out a black secretion. The Brown Hyaena wishing to leave its signature on the landscape seeks out a suitably tall, robust stalk of grass and straddles it. Crouching as it moves forward, the hyaena everts its pouch and, deftly wiggling its hips, inserts the chosen stalk in the groove between the grease-smeared lobes. In one smooth movement the hyaena steps forward and retracts the pouch, so drawing the stalk through the groove and then collapsing the black-coated walls around it. The grass springs free, coated with a 1-centimetre (0.4-inch) long blob of white secretion, about 70 centimetres (28 inches) above ground, and a slightly shorter blob of black secretion just over 1 centimetre (0.4 inch) further up.

These bicoloured signals are deposited roughly every 400 metres (quarter mile) within the territory, and at a higher rate near territorial borders because these are visited only rarely. The smell of the white grease is obvious to humans for well

over 30 days, whereas that of the black blob wafts away within hours, so the two tones of paste are probably conveying different messages. The short-acting black paste may inform other group members that an individual has recently foraged there, allowing them to avoid unproductive areas. The white paste seems more suited to communicating that 'trespassers will be persecuted'.

In the southern Kalahari, 45-kilogram (100-pound) Brown Hyaenas thrive on foods ranging from tsama melons through beetles to Wildebeest carcasses. However, they are mainly scavengers, gleaning a living cracking bones and slicing hide and sinew. So an individual is doomed when its sledgehammers wear flat, such as an old male that was seen trying for more than 30 minutes to splinter the leg bone of a Springbok that a youngster could have cracked in five minutes. Most Kalahari Browns live in clans and all forage in dry river beds where antelope, and hence carcasses, are most available. On average they travel about 7 kilometres (4 miles) between meals. Although their territories are larger where carrion is thinner on the ground, the number of adults in a clan does not depend on the size of the territory but on the size of the carcasses likely to be strewn around it. Sharing a Wildebeest with three to four companions is no disadvantage to a Brown Hyaena because it could not finish such a large carcass in one meal, and any left-overs would otherwise be taken by vultures and other scavengers. So clan size depends on the number of Wildebeest living or, more accurately, dying in each territory. In territories where Wildebeest are rare and smaller antelope are the mainstay, clans are correspondingly smaller.

A clan's den is hidden in remote sand dunes, perhaps out of the way of marauding Spotted Hyaenas and Lions. Female Brown Hyaenas generally produce a litter of three cubs every 20 months. Usually only the dominant female breeds, but if two litters are born in a clan the mothers will nurse each other's cubs, although favouring their own. The cubs are dependent for 12–15 months during which all the adults in the clan may carry lumps of carrion back to the den for them. One adult was observed dragging a 7-kilogram (15-pound) carcass over a distance of 15 kilometres (9 miles). This is not a paternal duty for clan males because they do not father cubs. Only subordinate males join clans, either their mother's or a neighbouring relative's, while dominant males roam through great tracts of arid land. When a female is on heat, one of these nomads will stroll into the territory as if he owns it, mate with her and then wander on. The resident males offer no resistance and often help rear the resulting cubs. Unlike the cuckolded male Aardwolves, clan males have not yet been seen to mate, and so their motivation to help seems to be mainly kinship. Those that are half-brothers to the new litter have a considerable stake in its well-being, and in easing the burden on their mother so that she may continue to produce more half-siblings for them. In Brown Hyaena clans studied in Botswana, resident

Brown Hyaena sampling a duo-tone paste mark deposited by a neighbour's anal pouch scent gland

males apparently feed only their half-siblings and not their cousins.

The Brown Hyaena's nearest relative, the 35-kilogram (75-pound) Striped Hyaena, is a shadowy creature of the night. It meanders alone in search of titbits in arid lands from North Africa to the Middle East. Sometimes two of these solitary wanderers meet in passing, but for the most part they communicate by scent, leaving perfumed messages for the next hyaena to read. So little is known of the Striped Hyaena that one can only speculate on its social life. Its diet seems similar in every respect to that of the Brown Hyaena, except that it rarely has access to big carcasses. So it probably has less inducement to form groups, and may live in pairs accompanied by subadult offspring. Some naturalists have suggested that the Aardwolf's zebra-striped coat mimics that of the vastly more powerful Striped Hyaena. This might fool enemies into leaving the puny Aardwolf alone, since few would attack a Striped Hyaena. However, although Striped Hyaenas did range further south until 1.5 million years ago, there are now none throughout the entire southern African range of the Aardwolf. Perhaps, instead, both these creatures have retained a primitive design once common to all hyaenas. Backing up this theory is the fact that the Brown Hyaena appears stripy in certain lights, resembling a darkened version of a Striped Hyaena. The design may originate even further back in the cat branch, because many civets and cats and even some mongooses are stripy. Such markings would doubtless have camouflaged early members of the cat branch in their original forest home. This, however, begs the question of why dog-branch Carnivores tend instead to be counter-shaded, darker above than below.

If there was once a stripy marking common to all hyaenas, the largest surviving member of the hyaena family subsequently changed its stripes into spots. At 80 kilograms (180 pounds) the Spotted Hyaena is the most powerful bone-crusher, able to tackle the largest carcasses, and this may be partly why it is also the most sociable of the hyaenas. Even living alone, a scavenging ancestral Spotted Hyaena would have needed a large territory to give it a fair chance of finding sufficient food. A large carcass would have been contested by mighty Lions, packs of Wild Dogs, pilfering jackals and innumerable vultures. As the burgeoning population of large ungulates provided more carcasses, allowing Spotted Hyaenas to feed side by side, the pressure of rivals probably forced them to operate in teams.

Due to their scavenging, ancestral Spotted Hyaenas were adapted, like dogs, to travel prodigious distances. They also had dual-purpose teeth, like *Pachycrocuta*, with sharp carnassials behind their sledgehammer teeth. This meant that there was no need for them to wait for ungulates to die. Like the pack-hunting dogs, the ancestral Spotteds had the stamina and build to run prey down on open ground, the collective strength to kill them, and yet were small enough to share the spoils. A modern Spotted Hyaena can chase an adult Wildebeest at 60 kilometres (40 miles) per hour for 5 kilometres (3 miles) before killing it. So the ancestral hunting hyaenas would have needed enlarged territories to ensure that having started a chase they could finish it on their own turf. Furthermore, the climate of the plains produced alternating rains and droughts, causing many antelope to migrate, and the hyaenas' territories would have had to encompass these movements too. All these factors would have pushed even a solitary ancestral Spotted Hyaena into occupying a vast territory that would often provide food for many companions.

The lifestyle of the Spotted Hyaena varies from place to place, confounding the popular image of the animal. They will rob a kill whenever their weight of numbers gives them the edge over any opposition. However, in the Serengeti plains of northern Tanzania, Spotted Hyaenas steal from Lions no more often than Lions steal from Spotteds. Indeed, modern Spotted Hyaenas are astoundingly versatile Carnivores that take full advantage of their dual-purpose dentition. They are active hunters and proficient scavengers. They can crack bones that would defy a Lion and can match the endurance of Wild Dogs. Their food supply enables, and encourages, them to live in clans whose size gives them a corporate power that is unrivalled, yet they are competent to forage alone. This versatility enables Spotted Hyaenas to live in very different places, such as the Ngorongoro crater of northern Tanzania, the nearby Serengeti plains and the Kalahari desert. These places offer similar antelope prey but under different circumstances, and the hyaenas have adopted correspondingly different lifestyles.

Confined within the walls of the 325-square-kilometre (125-square-mile) Ngorongoro crater, prey are abundant and non-migratory. There, Spotted Hyaenas live in clans averaging 55 members, each clan defending a 10–50-square-kilometre (4–20-square-mile) territory. Life is relatively easy, the hyaenas gaining the food they need in four hours of daily activity on average, during which they travel about 10 kilometres (6 miles). Territorial borders are marked with a cocktail of scents, including piles of faeces and anal pouch pastings, which ring the territory like the beads of a necklace. Fights with neighbours are ferocious, and the larger raiding party invariably wins. As is usual in Spotted society, all the adult females in the clan breed. The crater clans breed throughout the year and share a communal den, from which they set forth in hunting parties. They appear to plan the hunt in advance, big hunting parties going after big prey such as zebra, and smaller parties forming when smaller antelope are the intended victims.

In the arid environment of the Kalahari, the Spotteds hunt large antelope which are so thin on the ground that the hyaenas may travel 33 kilometres (20 miles) between meals in territories averaging over 1000 square kilometres (400 square miles). The clans number about eight members, probably because a Wildebeest or Gemsbok provides a square meal for a hunting party of eight. Each clan generally includes only one adult male and up to four breeding females. The territories are too vast to ring with latrines, so instead the Kalahari Spotteds pepper the interior with paste marks.

In the Serengeti, many ungulates migrate seasonally and some of the Spotted

Hyaenas there are nomadic. Others live in territorial clans on the long grass plains of the northern Serengeti. Each huge, 10- to 80-strong clan, with a slight excess of females, holds a territory of about 55 square kilometres (21 square miles) ringed diffusely with latrines. Prey move north to woodlands in the dry season and south to the plains in the wet season. The clan territories are positioned between the two, a compromise that produces adequate food within the territory for only three months of the year. The migrating Wildebeest herds pass southwards through the territories when the rains break in November, and then northwards again as the dry season bites in June. During these two periods the clans gorge. These Spotteds defend a territory of about 10 kilometres (6 miles) diameter, but why this size should be chosen over a larger or smaller territory is puzzling since it contains either a glut of prey or none, depending on the season. When prey pass through food is so abundant that the territory holders pay little heed to the nomadic Spotted Hyaenas which trespass in the wake of the herds. However, when itinerants arrive at a kill, the residents whoop to announce their priority of access.

After the Wildebeest have departed, bloated tummies soon grow hollow. The territory holders must cross their borders, travelling up to 40 kilometres (25 miles) on four-day sorties to the seasonal strongholds of their prey. A nursing female has to return regularly to her cubs, and may make 80 such trips a year, so commuting 6400 kilometres (4000 miles) annually. Meanwhile, cubs at the den go thirsty, and soon abandon the play that preoccupies them during the bountiful Wildebeest months. Such long intervals between meals may occur in other areas too, and probably explain why all Spotted Hyaenas have only one or two cubs in a litter. Each cub soaks up a prodigious quantity of milk when it gets the chance, one cub on record having hung unshakably to its mother's teat for five hours. So a female probably could not produce enough to quench the demands of a larger litter. However, in the Masai Mara, a northern extension of the Serengeti, prey are more consistently present and hyaena cubs have an easier life: they are weaned younger and mothers have litters at shorter intervals.

Spotted Hyaenas have also been studied in other places, including Ethiopia, where some are famous for scavenging in villages, and the Kruger Park of South Africa, where they live almost entirely by scavenging. The hyaena's shift from dedicated scavenger in the Kruger to hunter-scavenger in the Serengeti may be due to the ratio of Lions to Spotteds found in each. Lions are as numerous as Spotteds in the Kruger and their kills provide much to scavenge, whereas there are only half as many Lions as Spotteds in the Serengeti, forcing the hyaenas to do a lot of their own hunting.

While populations of Spotteds have adopted different ways due to the local food supply, there are common threads running through all their societies. For example, their sociality has necessitated a rich vocal repertoire, including a characteristic whoop audible over 5 kilometres (3 miles) or more. The whoops appear to convey

Spotted Hyaenas hunt cooperatively for Gemsbok in the Kalahari – Gemsbok without tails indicate a narrow escape from these professional hunters

much subtle information, those rallying clan members to hunt being responded to with a different level of urgency from those rallying members to drive off a Lion attacking their cubs. The whoops of adult males, females and cubs are different, as are those of each individual. A mother returning from a long hunt and her thirsty cubs, which may have wandered several kilometres from the den, will swiftly find each other by recognising each other's whoops. Similarly, adolescents beginning to follow the hunt easily get lost, but their mother can summon them to the kill by whooping. Hyaenas also whoop to show off as individuals, the rate and style of whooping indicating social status. Because of this hyaenas whoop singly, rather than in chorus as do wolf packs advertising their corporate strength. Males whoop more than females of comparable rank, but the dominant female often engages in the longest bouts of whooping of all. Blasting out these calls may be a genuine reflection of an individual's stamina, since bellowing is an energetic activity.

As they mill around a carcass a scrum of Spotted Hyaenas may appear chaotic, particularly after they have travelled for five days to find a meal. However hungry they are, though, there is a strict hierarchy and females reign supreme. Although their skeletons are no larger, females are 12 per cent heavier than males and dominate them (male Brown and Striped Hyaenas are slightly heavier than females).

The hierarchy is a nepotistic one, the offspring of high-ranking mothers taking precedence at food over all lower-ranking animals, including adult females subordinate to their mother. This head start provides a lifelong advantage, the daughters of high-status mothers eventually inheriting high status themselves. Not surprisingly, low-class lineages do not relish undertaking the rigours of a perilous hunt with a mob that then bars them from the spoils. So when clans split up to hunt, and when they split permanently, they tend to do so along maternal lineages or matrilines.

Females' domination at kills may be a necessity if they are to secure enough food to generate the milk that their cubs need. In some habitats, such as the Serengeti and Kalahari, Spotted Hyaenas often feed too far from their den to drag food back. Regurgitation may be ruled out by the speed at which flesh is processed in the acidic cauldrons of their stomachs. In any event, the option selected by female hyaenas is to gorge and produce copious milk in huge udders that can carry 3 to 4 kilograms (6.5 to 9 pounds). Hunting hyaena-style requires strength and endurance, so the cubs must depend entirely on milk for nine months and cannot contribute to the hunt until 16 months old.

The females' need to dominate at the kill has consequences that reverberate throughout Spotted Hyaenas' lives. According to African folklore, these animals are hermaphrodites – that is, each individual is both male and female – because the females appear to be equipped with a perfect set of male genitals. In fact, the female's clitoris is enlarged to look like a penis, distinguished only by being slightly blunter. Furthermore the lips of her genitals are fused together to form a sac in which fatty lumps mimic testicles. When two females greet, both groan and raise a leg so that each may inspect the other's erection. The greeting initially involves sniffing each other's genitals, a compromising position in which both parties are exceedingly vulnerable and which may build trust between them. The two then sniff and lick each other's genitals for 10 to 15 seconds, amidst much vocalisation.

Many scientists have sought an explanation for the female Spotted Hyaenas' genitals. The first suggestion, published in 1877, was that they had developed so that members of the same sex could enjoy sexual gratification with each other. The most recent suggestion is that they derive from the lactating female's intense need for meat and so for dominance, which can be secured through strength and aggression. Physiologically, the easiest way to tweak the body's chemistry to increase muscular bulk and aggression is to increase the amount of male hormones, androgens, which circulate at lower levels in all female mammals. Masculine genitals would be a side effect of this physiological change, as in a sad case some time ago when a drug was prescribed to pregnant women threatening to miscarry. It acted in the same way as androgens and girl babies of these women were masculinised just like female hyaenas. Among female hyaenas, too, masculinisation may have been an accidental side effect but, once started, it is easy to see how the process

Female Spotted Hyaenas sniffing each other's erect genitals and imitation scrotal sacs. These masculine qualities are associated with unusual features of female hyaenas' sex hormones that have astounding consequences for their social lives

could gain in importance. Perhaps if male-like features became part of female displays indicating dominance or subordinance, females would develop ever more exaggerated male-like organs. Certainly, today Spotted females have the same level of androgens in their blood as males (although they are slightly different compounds), and this affects their young as well as themselves. The aggression and strength-promoting effects of androgens in all mammals are most pronounced when a youngster is exposed to them during critical periods in the womb. During the last third of pregnancy, the mother hyaena's ovary produces a massive dose of androgen that deluges her unborn young.

Studies were made of 99 litters (57 of which had twin cubs and 42 single) at the communal den of one large clan of Spotted Hyaenas in Kenya's Masai Mara reserve. These studies revealed that both twin brothers and twin sisters were surprisingly uncommon. Since overall, Spotted Hyaenas have equal numbers of sons and daughters 50 per cent of twins should have been the same sex, but only 15 per cent of them were. A clue to the puzzle came with the first observations of Spotted Hyaena births. A female gives birth through the 1.5-centimetre (0.6-inch) long opening at the end of her 'penis'. Having been born by a process that looks painfully akin to being squeezed through a pea-shooter, the newborn emerge ferocious. Their birth must be even more problematic because relative to their mother's weight, they are the biggest of all Carnivore babies. Unique amongst mammals, these infants are born with their eyes open, and with well developed 6–7-millimetre (about 0.25-inch) long milk canine teeth, 4-millimetre (0.15-inch) long incisors, and the muscular coordination to perform a killing neck-shake. They are also awash with aggression-stimulating male hormones, the female cubs as much as their brothers. Still slimy from the business of birth, the first born will stumble manically towards a sibling, grasp it on the nape of the neck and shake for all it is worth. As soon as the second born has caught its breath it retaliates with similar fervour. In one case of triplets, the first two were seen to fall upon the last born even before it had emerged from its birth membranes.

This cradle violence is so fierce that a quarter of all cubs die of starvation, prevented from nursing by bullying siblings. The worst fighting is between twins of the same sex, which could explain the strangely distorted sex ratio of litters in the Masai Mara clan. Probably half the litters are same sex twins at birth, and in most of these cases one sibling kills the other. The reason for the particular ferocity between same sex twins is unknown, but it may be linked to the benefits that could accrue in later life. In hyaena society, twin females especially are destined to be lifelong rivals; so removing one at the outset would give the other a tremendous advantage. In contrast, there is likely to be little competition between brother-sister pairs once they mature, because males leave the clan after puberty. In another twist to the plot, recent evidence indicates that high-ranking females specialise in producing singleton sons, and perhaps they achieve this by partisan refereeing of the infantile brawling.

There is a fundamental difference of interest between a mother and her cubs in all mammals. The mother's aim is to rear as many healthy offspring as possible, while each cub gives top priority to its own selfish concerns. Spotted Hyaena breeding dens are so constructed that adults cannot fit inside. This probably stops raiders from neighbouring clans or rival matrilines eating the cubs, but it also stops the mother acting as a referee to their squabbles. The infantile aggression is worst for the day or so after birth, but may continue once the cubs emerge from the den when the mother struggles to keep the peace, repeatedly pulling the youngsters apart.

In the 60-strong Serengeti clans as many as ten nursing mothers may share a communal den. This may be another reason why, unlike Brown Hyaenas, Spotteds do not drag carcasses back to the den. Among a dozen or more squabbling cubs the smallest would lose out. Spotted Hyaena mothers very rarely suckle another female's cubs, so the young have only one competitor at most for their mother's milk. At the centre of the huge clan is a single dominant female. She rests at the prime site near the centre of the den, while lower-ranking mothers move subserviently on the perimeter. Dominant females also produce most milk and have cubs at shorter intervals than do subordinate ones. So, over the years, they churn out more descendants.

Clan members come and go from the communal den alone or in small hunting parties. However, with so many members there is almost always somebody at the den to raise the alarm if danger threatens. Lions and nomadic hyaenas are both known to raid communal dens and kill the young. Females guarding cubs whoop more than those without, perhaps signalling their presence as a deterrent to such infanticidal interlopers. As well as facilitating round-the-clock vigilance over the cubs, communal denning also provides a meeting place where cubs can get to know adults and learn their place in society, and where adults can exchange information on good places to forage.

Spotted Hyaena society offers different rewards to the young males and females that survive the gauntlet and reach puberty. Young females are generally recruited into their mother's clan, where their status depends on hers, while young males tend to emigrate. Some of these Spotted males become nomadic like their Brown cousins, but the best-quality Spotted males immigrate into another clan where they sire most of the cubs. There is some evidence that the sons of dominant mothers have unusually high levels of male sex hormones and are therefore especially aggressive in the battle for breeder status.

Despite this, an immigrant male advances timorously on a female and may even be chased away by her young daughter. He may select a clan after listening to the whoops of the current residents, from which he can judge their number and age. He will then have to work hard to secure a place. It often takes more than a year of lurking on the periphery, being chased on sight, before a successful male can insinuate himself into a clan. One would-be immigrant in the Kalahari spent 27 months failing to be accepted, but was eventually rewarded when a subordinate matriline split off from the clan to join him. Once admitted, an immigrant male may

(Overleaf) Infant Spotted Hyaenas fight furiously within moments of birth, aggression that often ends in siblicide. The saliva-soaked napes of the necks of these two cubs are hallmarks of attempts to bite each other

lead hunting expeditions but at the kill he must quickly bolt food before being forced aside by the matriarchs.

On one occasion, a male was seen to circumvent the usual admissions procedure. The nomadic male tried to join a clan in Botswana's Okavango delta and was, predictably, driven away. Rather than giving up, the nomad forged a remarkable alliance with the number two male in the clan hierarchy. Having trailed the clan for two days, scent-marking as he went, the nomad again challenged the dominant resident. This time his new ally, the number two, came to his aid, and soon the loyalty of the whole clan switched. They fell on the old dominant and drove him off. The number two male then ascended to the dominant position and the kingmaker entered the hierarchy directly beneath him.

An immigrant that secures membership of a big clan faces a long climb up the social ladder. A handful of dominant males monopolise almost all matings with the 20 or so breeding females and dominance appears to be a reward of long service. This is not surprising, since in a harsh world, survival is proof of high quality. The dominant males of large clans are old and have devoted their adult lives to forming affiliations with hostile females, including young females that they befriended as cubs. One way they do this is to whoop a lot, keeping the sound of their personal voices in the females' ears. The dominant males are also distinguished by whooping in the most energy-consuming way, perhaps to show their prowess both to females and rival males. When a male whoops a close rival may reply in what amounts, literally, to a shouting match. Recent immigrants, despite their lowly status, whoop a lot, perhaps to establish their voices in the female's thoughts.

The fiction that witches ride Spotted Hyaenas seems modest beside the reality of females with fatty imitation testicles grooming their androgenous daughters through favouritism for high status. The reality may turn out to be stranger still because the society of Spotted Hyaenas has an intricacy far beyond our knowledge. In building their society, they have added strength of numbers to the power and versatility of their jaws to become the most abundant large Carnivores in Africa today. Yet 15 million years ago, in the heyday of the hyaenas, the Spotted and other modern hyaenas would have appeared aberrations. While mainstream hyaenas were following a dog-like opportunist trade, any that turned to a bone-cracking or termite-eating trade would have seemed rank outsiders in the lottery of evolution. But circumstances changed, and for the last few million years at least, their eccentricities have paid off.

Family Tree: Hyaenas

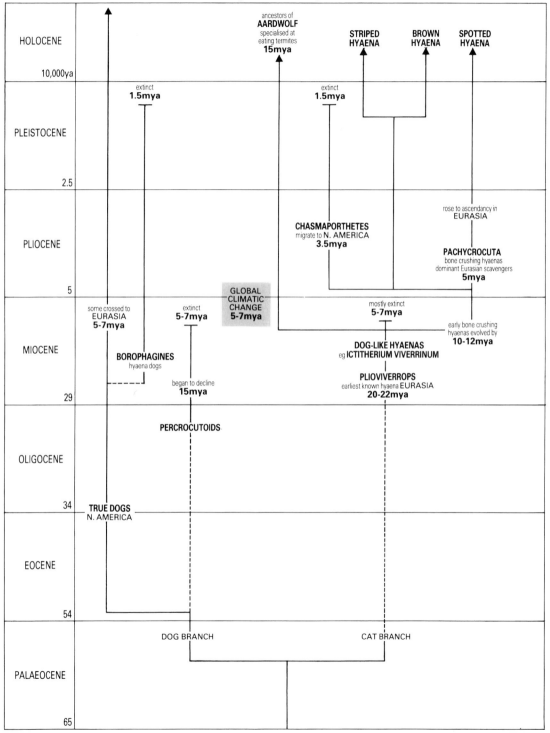

CHAPTER 5

LIFE IN THE SLOW LANE

In 1869 the first Red Panda was brought from the Himalayas to London Zoo. The size of a spaniel, with a look of surprised innocence on its white-muzzled face, this otherwise reddish-brown fluffy creature would have reminded zoo visitors of a teddy bear, but for the fact that such toys were not invented (and named after President Theodore Roosevelt) until 33 years later. Having been identified as a Carnivore from its teeth the Red Panda, which occurs naturally from Nepal through Burma to southern China, was plied with prime meat, which it refused. At first this seemed an understandable reaction to the tribulations of its journey. However, it continued to reject food, lapping unenthusiastically at sweetened beef tea, egg yolk and porridge, until its life hung in the balance. At this point the zoo's Superintendent, Papa Bartlett, took it on a leash for a walk through the gardens. The Red Panda paused at a rose bush, deftly picked a bud and ate it. A few paces later it feasted greedily on the yellow fruit of the bush *Pyrus vestita*, a species of pear. Then realisation dawned: the new Carnivore was a vegetarian!

In 1937 the first Giant Panda followed its red namesake west to London Zoo. The two had a lot in common, including a vegetarian diet, similar teeth and skulls, and the uncommon trait of a short, backward-pointing, S-shaped penis. Most distinctively, both had a 'thumb' which was actually a wrist bone remarkably reshaped into a sixth digit to grasp the bamboo both relished. Because the pandas reminded the French biologist, Cuvier, of cats, they were given scientific names incorporating the Greek word for cat, *Ailurus*. However, further investigation revealed closer links with the dog branch of the Carnivores, particularly the bears (Ursidae) and the raccoons (Procyonidae). While no other members of these families are such strict vegetarians, most are omnivores and partake regularly of vegetation.

The bears, raccoons and pandas descend from Carnivores like *Vulpavus*, tree-top hunters that lived in North America 40 million years ago. Some vulpavines adapted to life on the ground and provided the stock of the dog-branch of the Carnivores (see page 34). The families they gave rise to about 35 million years ago are distinguished partly by the number of molar teeth they have on each side: dogs have three upper and three lower molars, bears have two upper and three lower, procyonids have two of each, and the mustelids have one upper and two lower molars. While the true dogs stayed in America for millions of years, other dog-branch Carnivores soon migrated to the lush forests of Eurasia which became the centre for their early diversification. Amongst the creatures that flourished in Eurasia was a family of stocky yet fleet animals that resembled bear-like dogs and so are called the bear-dogs or half-dogs (Amphicyonidae, see p. 34).

North American fossil half-dogs were rarer and less diverse than their Eurasian counterparts, possibly because of competition with true dogs there. Nonetheless, between 25 and 15 million years ago they were the dominant dog-like Carnivores of the northern hemisphere. One, *Daphoenodon superbus*, was 2 metres (6.5 feet) from nose to tail-tip, with heavy jowls and a wolfish demeanour. In the Nebrascan floodplain, a region of rather dry, open savanna, a family of *D. superbus* became immortalised as fossils one wet day 20 million years ago. A river running through the plain rose in a flash flood and drowned the family in their den. The sediments later hardened into rock, preserving the remains of one young adult female and a juvenile, and providing the earliest record of Carnivores living in dens. Later fossils suggest that some bear-dogs lived socially in large burrows dug high in river banks. Since these creatures also had sharp blades to their scissor teeth, like the modern African Wild Dogs, it is possible that they too ran down prey as a pack.

Other bear-dogs opted for a hunter-gatherer omnivorous lifestyle, doubtless to escape competition from specialist meat-eating Carnivores. Omnivorous bear-dogs flourished for a while but then came an increase in the numbers of ground-dwelling competitors: dogs in the New World, dog-like hyaenas in the Old World, and bears and procyonids in both. Bear-dogs may have initially blocked the evolution of these rivals. If so, the tables were turned 15 million years ago when the boom in dog-like hyaenas in Eurasia and the rise of the true dogs in America coincided with the decline of the bear-dogs. Since the early dogs, like *Hesperocyon*, were no bigger than a small civet it probably took a while before they evolved sufficient size to compete with bear-dogs. Eventually the bear-dogs, a compromise in design as well as name, were outweighed by bears as well as out-paced by dogs, and became extinct three million years ago.

The bear-dogs might have survived when they came under pressure 15 million years ago if they had been able to move further towards the vegetarian end of omnivory. However, by then the procyonids and bears had beaten them to it. These related families had split from the dog branch about 35 million years ago. At that time, members of both families had been mainly meat-eaters, but in the intervening millennia they had specialised increasingly on plant foods. The bears opted for a large size, giving them a life in the slow lane sustained mainly by fruits and nuts.

(Overleaf) Daphoenodon superbus *was a half-dog or amphicyonid living in North America 20 million years ago*

The procyonids also developed a taste for a mixed diet but, being generally smaller, some opted for a frenetic, opportunistic lifestyle. In contrast to omnivorous dogs, most members of both families climbed trees as well as foraging on the ground.

Among modern procyonids, the Cacomistle of the dry forests of Central America most closely resembles the earliest procyonids in its climbing skills and carnivorous diet. Its tawny body and dainty white paws are attached to a flamboyantly luxurious black-and-white ringed tail. To judge by the air of prim disapproval in its huge white-rimmed eyes, the Cacomistle appears to be shocked by its incongruous attire.

The Ringtail is closely akin to the Cacomistle but slightly smaller, with more rounded ears. It lives in arid lands from Texas, north-west to Oregon and south through Mexico. Both species are very agile, and use their long tails for balance. They can rotate their hind feet by at least 180 degrees and this, together with their high-traction footpads and, in the case of the Ringtail, retractile claws, enables them rapidly to descend steep cliffs and tree trunks head first. Ringtails kill efficiently by pouncing on their prey and biting the nape of its neck. Like Cacomistles, they feed on rats, mice, squirrels, birds, reptiles and fruit (being especially fond of persimmons), but their mainstays are crickets and grasshoppers. Such small prey does not foster collaboration, and for the most part Ringtails are seen alone. However, there are reports of fathers bringing food to young, and they reputedly live in pairs within territories of about 100 hectares (250 acres).

The Cacomistle and Ringtail are the only members of their family that remain sufficiently predatory to have sharp scissor teeth. Their carnassials have lost only a little of their ancestral sharpness through their flirtation with omnivory. Most others have opted for a broader diet, and replaced the scissors with multi-purpose grinding teeth. The most notorious is the 5–10-kilogram (11–22-pound) Common Raccoon which wears a black 'bandit's' mask. This procyonid has reached the pinnacle of opportunism eating anything and living anywhere, including household attics, where more than a dozen may cohabit. It has a habit of rummaging under-water, which was thought to be an attempt to wash food, but is probably to uncover it. Common Raccoons are highly dexterous, with sensitive front paws with which they feel underwater for prey such as crayfish. They also eat anything from frogs to fruit, eggs, nuts and maize, which makes them unpopular with farmers. This, together with their valuable fur, may motivate the hunting that crops 4 million raccoons annually.

Raccoons are polygynous, one male occupying a territory that embraces two or three females whose ranges overlap widely. The females, generally mother and daughters, forage alone but often den together. However, in parts of Washington City there are 80 raccoons per square kilometre (200 per square mile), and at a

Ringtails of southern USA look and, probably, live much like the earliest Procyonids

restaurant in Florida where they are fed as a tourist attraction up to 50 may congregate. The social ties among these urban dwellers are unknown.

The Common Raccoon's affinity for water probably helped it give rise to five Caribbean island species: the Barbados, Bahama, Guadaloupe, Cozumel and Tres Marias Raccoons. Their ancestors may have been transported to the islands by people, but earlier procyonids undertook similar journeys by clambering aboard rafts of vegetation. Five million years ago, when there was no landbridge between North and South America, they drifted south to become the first true Carnivores to colonise South America. The castaways arrived on the shores of what was to become Ecuador, and prospered in a land of thunder-birds and pouched killers (see pages 14–40). In only a few hundred thousand years they diversified and some grew huge, becoming the 'bears' of the south. However, their success came to a jarring halt two million years ago when Central America collided with South America. North American Carnivores, including true bears, swept over the landbridge and annihilated the bear-like raccoons (see page 36).

(Left) Raccoons feast on the autumn berry crop

A super-abundant food supply on the refuse of a Florida restaurant (below) has major consequences for raccoon social lives

Today's South American procyonids descend from later land migrants rather than from the early rafters. They include a more arboreal version of the northern raccoons, the Crab-eating Raccoon, and two species of coati. One of the coatis lives along with the Crab-eating Raccoon, on Mexico's Cozumel island, while the other, the White-nosed Coati, occurs throughout much of South America. Coatis are 4-kilogram (9-pound) forest-dwellers that climb well and nest in trees. Both species have a long counter-balancing tail and reversible ankles to aid in head-first descents. Aided by long, whiffling noses they forage on the soggy forest floor for insects, supplemented by frogs, lizards and their eggs, and small mammals.

Female and young White-nosed Coatis travel as a stable band numbering up to 25, including up to seven adult females, in an 80-hectare (200-acre) range. They share in looking out for danger and females cooperate in nursing young, apparently irrespective of kinship. Adult males of three years or older tend to travel alone. Having nobody to groom them, males have more parasites than band members. They are also 20 per cent heavier than females and may be attacked if they approach the band, which led to them being regarded mistakenly as a separate species, the Coatimundi (Guarani Indian for 'lone coati'). Outside the January-March mating season, males generally live alone in overlapping ranges, although sometimes two males travel amicably in company, and sometimes they join female bands. Many have deep wounds and broken teeth from fierce fights. During the breeding season, a male inveigles his way into the female band by grooming its members and behaving submissively, and then mates with all the females, in trees. Soon afterwards, the band expels the male, and the females disperse to build tree nests in which three to five young are born. One reason that females attack males is that they sometimes kill the young. However, later, when the band re-forms, the father may briefly join them and groom the juveniles. This may help him recognise, and avoid accidentally killing them, in later encounters.

Amongst the procyonids that have moved furthest away from a meat diet are the Olingo and Kinkajou, the most ancient division of the family. Both are nocturnal, weigh about 2 kilograms (4.5 pounds) and live in Central and South American forests. Indeed, on a superficial glance, they can only be distinguished by their tails, the Olingo's being long, bushy, slightly ringed and not prehensile, and the Kinkajou's shorter, plain and prehensile. The Olingo supplements a diet of fruit and nectar with small mammals, birds and insects. This places it mid-way between a conventional procyonid and the extremist Kinkajou, which lives almost exclusively on fruit, nectar and honey and has an especially long tongue for probing flowers' nectaries. Both species have short faces and large eyes, possibly to aid nocturnal tree-climbing, and have converted their scissor teeth to flat fruit-crushers. Their canines have

Raccoons are the ultimate opportunists, adapting to a wide variety of diets. Sifting the mud in shallow water they are adept fishermen

pinstripe grooves, a feature shared with the Small-toothed Palm Civet (see page 30), which is also a fruit-eater. Kinkajous also differ from other procyonids in having only one young annually, one less premolar tooth, and no anal scent gland; instead, they have a scent gland on their chest and belly. Olingos often forage in pairs, sometimes in the company of several Kinkajous, while Kinkajous may forage in large groups.

(Left) Coatis forage by day in large troops in the forests of Central and South America

The Kinkajou's long tongue (below) is adapted to probing forest flowers for nectar

The procyonids span the spectrum of omnivory, with social lives to match. The nocturnal, largely insectivorus Cacomistle has a diet similar to that of arid-land foxes and seems, like small foxes, to live as territorial pairs. The Common Raccoon is more omnivorous, taking more fruits and carrion, which is a more variable food supply and one which sometimes provides a bonanza. So, more individuals, although generally not more adult males, can be squeezed into the territory wherein each continues to forage alone by night. Where food is superabundant, super-groups may develop. The Coatis forage by day mainly for insects and have to watch out for eagles and other predators. Since their insect prey are soon replenished after they have harvested an area, it costs them little to expand their territories to fit in a larger band and so have many pairs of eyes to keep a lookout. Even a big band fits into a relatively small range. The Olingo also eats insects but, being nocturnal, would benefit less from many lookouts. Moreover, the rodent and insect components of its food are less readily shared, so it lives in pairs. However, when a tree fruits in an Olingo pair's territory they tolerate the company of a band of Kinkajou. Whether the two species join forces when threatened by a predator is unknown. The Kinkajous' heavy dependence on fruit allows them to congregate because a tree in fruit provides food for plenty and the fruit will only spoil if left uneaten. In tropical forest, trees fruit year-round, although only one species may be in fruit at any one time. Few details are known of the Kinkajou's social life, but perhaps a group holds a territory big enough to ensure that there will usually be at least one tree in fruit within it. When there is not, they may split up and forage in smaller groups or alone.

The other family that relies heavily on fruit and vegetables, the bears, has developed differently from the procyonids, and its members are now gigantic in size and occur predominantly in cold or temperate climates rather than the tropics. Fossils of the earliest bear, the wolfhound-sized *Cephalogale*, come from 37-million-year-old Chinese deposits. However, the first known member of the lineage leading to modern or ursine bears was *Ursavus elmensis*, which arose more than 20 million years ago and was probably rather like today's Raccoon Dog (see page 83), an omnivore from Asia. Measuring 75 centimetres (30 inches) at the shoulder, *Ursavus* was dog-sized and probably an adept climber, equally at home in the trees and on the ground. To judge by its broad-topped cheek teeth, it supplemented flesh with a mixed and chewy diet of fruits and tubers. Over time, bears became bigger and the length and crushing surface of their molars increased while the blades of their scissor teeth diminished. As heavyweights it benefited them to walk on the soles of their feet, rather than on their toes like dogs.

Some of the early bears may have played a part in the evolution of sealions and walruses and, possibly, the seals. Fossils indicate that these carnivorous marine mammals first appeared about 20 million years ago. Studies of a blood protein, albumen, in modern sealions, walruses and seals confirmed the anatomical evidence from fossils that they are descended from dog-branch Carnivores. However, molecular biologists tend to conclude that all these marine mammals had a common origin, whereas anatomists favour the idea that bears gave rise to the sealions and walruses, while the weasel family (a sister family of the procyonids) gave rise to the seals.

Shortly before these marine mammals arose *Ursavus elmensis* spawned the Asian lineage of early ursine bears that probably lived much like the Sun Bears of modern Southeast Asia. At 27–65 kilograms (60–140 pounds) they are the smallest contemporary bears, and get their name from an orange chest mark on their otherwise mainly dark fur. Sun Bears nest in trees, and climb well using curved claws, large paws and naked pads. They feed on fruit, succulent growing tips, insects, small mammals and birds, and males are 20 per cent heavier than females.

Most early bears probably prospered on a similar mixed diet, but some may have specialised, as the Sloth Bear does today in Asian forests. The 100-kilogram (220-pound) Sloth Bear is so named because it can hang upside down, sloth-like, from branches. It specialises on termites and so is to bears what the Aardwolf is to hyaenas and the Bat-eared Fox is to dogs. The tools of its trade include massive 8-centimetre (3-inch) long claws to rip termite mounds asunder, and a long snout which houses a cavernous palate and an outrageously long tongue. Once the mound's fortifications have been breached, the bear contorts its naked, flexible lips and long tongue into a tube. It puffs out to blow the dust off the termites, and then closes its nostrils and inhales gustily to suck the insects up like a shaggy vacuum cleaner. Sloth Bears have no upper incisor teeth to obstruct this suction tube. Only a few thousand Sloth Bears survive today, harassed by timber cutting, but benefiting from parks set up to protect the Tiger.

Mainstream bears gradually became more omnivorous, taking more plant food in their Asian forest home. At the same time the world was becoming drier, and humid forests were replaced by temperate forests and scrubland. Between twelve and ten million years ago the bears spawned a new offshoot that could capitalise on the open spaces being created: the great running bears or tremarctine bears. These moved away from forest omnivory, and some became giant, carnivorous sprinters. They dispersed across huge distances, eventually crossing to North America.

The running bears' success peaked during the Pleistocene, from 1.8 to 0.2 million years ago. They included the 600-kilogram (1300-pound) Short-faced Bear, which was rangy, long-limbed, fast-moving and rapacious. Another, the Florida Cave Bear, became a 400-kilogram (900-pound) vegetarian. In both species, males were twice the weight of females. When the Panamanian landbridge surfaced two million years ago, it was running bears that travelled south to seal the fate of the bear-like raccoons. The Short-faced Bear ranged over both North and South America until it became extinct, along with other running bears, about 10,000 years ago. The large,

carnivorous running bears may have suffered from competition with the big cats, but it was probably the late Pleistocene extinction of their large herbivore prey that wiped them out along with Dire Wolves and North American Lions. Only one running bear, the Spectacled Bear of South America, lives on, albeit much modified, as the sole survivor of a 10–12-million-year-old lineage. An accomplished opportunist, it lives largely on shoots and fruits, but will take insects, carrion and young deer.

While the running bears were diversifying and spreading to North America, their Eurasian counterparts were becoming more ponderous and omnivorous. These Eurasian bears were not much perturbed by the climatic changes seven to five million years ago that so disrupted the lives of hyaenas, dogs and other families. About five million years ago, the Little Bear appeared. This was to become ancestor to six of the eight bears alive today: the Sun, Sloth, American and Asian Blacks, Brown and Polar Bears. The Little Bear was Sun Bear-sized, but its canines were much sharper and its scissor teeth were still slicers, although less fiercely so than those of its predecessor, *Ursavus*. By 2.5 million years ago Little Bears had grown larger, possibly to cope with the advancing Ice Age because large bodies lose heat

(Left) Sloth Bear lips are adapted to form suction tubes for 'hoovering' up termites

The Short-faced Running Bears (below) descended from the tremarctine lineage that migrated to North America

more slowly than small ones. The larger bears gave rise to the Etruscan Bear, which still retained the forward premolar teeth that are lacking in its descendants. At first it was the size of a smallish Black Bear, but by the time of its extinction in an interglacial period 1.5 million years ago, it was as big as a Brown Bear.

The Etruscan Bear produced three lines. One, in Europe, led to the massive Cave Bears which lived from about 500,000 to 10,000 years ago. At 400 kilograms (900 pounds), they were the most powerful Eurasian Carnivores of the day, three times the size of a modern European Brown Bear, with disproportionately massive heads. Despite this, Cave Bears were vegetarians, like their similarly sized New World relatives, the Florida Cave Bears. The Eurasian Cave Bears probably encountered early humans who sought to share their homes, and the Brown Bears certainly did. The Trois Frères cave in Ariège, France, contains an engraving of a vomiting Brown Bear bristling with spears. Some European caves contain passages with polished walls, *barenschliffe*, rubbed smooth by the passing bodies of bears over hundreds of thousands of years. The remains of 50,000 Cave Bears have been found in deposits from the Dragon's Cave at Mixnitz, Austria, and these illustrate some of the hazards of interpreting fossils. The fossil bodies vary in size and this was at first taken as evidence that there were giant and dwarf bears and that the dwarfs predominated latterly. In fact, the giants were males and the dwarfs females, the latter being found in smaller caves because these were the ones used largely by females. To further confuse matters, museums kept the 'best' or largest specimens, with the result that 90 per cent of museum Cave Bears are males. Despite such bias, it does seem that Cave Bears, like Wolverines and hyaenas, grew larger during glaciations and smaller during interglacials, again probably to adjust the rate of heat loss.

The other two lineages that descended from the Etruscan Bear grew up in Asia and led to the Brown and Black Bears. At least 1.5 million years ago, members of both these species had padded across the Bering landbridge to North America. As ghoulish evidence of their arrival, many were trapped in the Californian La Brea tar pits alongside the corpses of Short-faced Bears.

Today, bears are found around the world, excepting Australia, Antarctica and Africa. Traces of only one species have been found south of the Sahara, and this left abundant fossils on the southern Cape in South Africa. How it got there without leaving any more northerly signs is a mystery. There are now eight species of bear worldwide; there have always been only a few species at any one time. This may be because the bears' success has been based on a catholic diet, inhibiting the development of specialists. Their varied tastes also meant that bears could have huge geographical distributions, although these have been decimated by people. The Brown Bear, for example, once spanned the northern hemisphere but is now scarce in Europe outside European Russia, and much reduced in North America.

The running bears were daunting predators, but it is unlikely that even they could cope with the mace-like tail of a Glyptodon

Amongst the largest of today's bears, Brown Bears range in size from big to enormous, depending partly on the abundance of their food. There are two North American subspecies: the Grizzly and the Kodiak. Grizzlies inhabit the inland forests, and most weigh between 160 and 315 kilograms (350 and 700 pounds), the heftiest being about 450 kilograms (990 pounds). Kodiaks are restricted to Alaska's Kodiak Islands, which were cut off from the mainland 10,000 years ago. There are now between 2500 and 3000 Kodiaks, males being 680–815 kilograms (1500–1800 pounds), with a 3.4-metre (11.1-feet) tall record holder weighing 1000 kilograms (2200 pounds). Their bulk breaks the general rule that island races are smaller than their mainland relatives, perhaps because their large size helps heat conservation in the far north. These giants are protected by their tourist value, each Kodiak Bear being worth US$10,000 a year to the islands' tourist business. Other contenders for biggest Brown Bear include the Siberian Brown Bear, which can weigh 816 kilograms (1795 pounds), and there is an unsubstantiated report of a 1130-kilogram (2490-pound) individual from Kamchatka in eastern Russia. In comparison, males of the Italian and Spanish subspecies are midgets at only 80–300 kilograms (180–660 pounds).

Big size means big appetites and sometimes long journeys to find food. For example, female Brown Bears in northern inland parts of America, where food is scarce, may have to cover areas of up to 200 square kilometres (80 square miles) to satisfy their hunger. Brown Bears can run down elk, sometimes over a chase of several hundred metres. However, generally they are compelled by their bulk to rely on static plant foods. Their size also affects the rate of their growth and reproduction. Brown Bears do not mature until they are between eight and ten years old and only produce one or two cubs every four to five years.

In all species of bear, the mother nurses the cubs unaided and the male concentrates on fathering as many offspring as possible, which necessitates defeating other suitors. This has resulted in males becoming larger than females. The males are now vast fighting machines, scouring huge areas in search of females. In the region where female Brown Bears occupy ranges of 200 square kilometres (80 square miles), males cover 1100 square kilometres (420 square miles) and may weigh 80 per cent more. Male Asian Black Bears are up to 70 per cent heavier than females, while male Spectacled Bears can be triple the weight of females.

Most knowledge of bear society comes from a study of Black Bears in Minnesota, USA. Black Bears occur from New Mexico to Alaska and east to Newfoundland. In the east, where acorns and beechmast abound, females average 90 kilograms (200 pounds); to the west they average 65 kilograms (145 pounds). Males are up to 50 per cent heavier, the record being 272 kilograms (600 pounds). Females in the eastern acorn belt start breeding between three and five years, producing two to four cubs every two years. Those further west first breed between four to eight years, having 1.7 cubs every two to four years. In Minnesota some depend on clumped food, for example garbage or spawning salmon, and these tend to congregate and a pecking order develops among them. More commonly, food is widely scattered, and female Black Bears hold 10-square-kilometre (4-square-mile) territories. Males, which are one third heavier, quest for females over more than 100 square kilometres (40 square miles). Males' ranges are too large to defend, so they overlap. Females appear to have little choice in their sexual partner, being mated by whichever male is victorious. However, they try to increase the number of candidates by wafting aphrodisiac urinary perfumes around well before they come on heat. This ensures that lots of suitors are lured to the area, the 'best' of which will win. A litter may have more than one father because each female comes into heat several times during the breeding season. Furthermore a female Brown Bear was seen to mate ten times with four mates in two hours. It is rumoured that females bearing one cub may abandon it, doing better to breed again with a larger litter the following year.

At about a year and a half old, Black Bears leave their mother and begin a solitary life. The mother smooths this transition by relinquishing portions of her territory to the yearlings. She avoids these enclaves while the youngsters secure their foothold in the property market. Eventually the sons leave home altogether, but the daughters gradually increase the size of their enclave until they mature at four years old. Some mothers even vacate their territory in favour of their daughters, an inheritance that gives them a great head start. A less privileged young female has to fight for territory and may end up wounded, pregnant and homeless, giving birth in a den within a hostile female's territory, before setting off on a vagrant life with her youngsters.

Choosing whether to evict her daughters or abdicate in their favour must be difficult for a mother Black Bear. In theory, the choice should depend on whether the mother will gain more descendants in the long run by cosseting her daughters (and so increasing the number of her grandoffspring) or looking after herself (and so producing more offspring and, possibly, grandoffspring). This will depend in turn on the food supplies in her territory, her strength, age and likely reproductive future, and the quantity and quality of her daughters and the likelihood of them finding territories elsewhere. Of course, a mother bear will not coldly weigh up all these factors but, like all animals, she has been fashioned by natural selection to behave in ways that maximise her descendants.

Most female Minnesotan Black Bears stay in or near their mother's territory so presumably the food supplies within the territory are ample or space elsewhere is scarce. Between males, though, there is an unbridgeable generation gap: old males will eat any young males they can catch. Nobody knows whether they spare their sons, or merely assume that male offspring will have dispersed. Young males leave home by their fourth birthday, generally settling at least 50 kilometres (30 miles) away. They have been known to disperse as far as 219 kilometres (136 miles). Their departure does not coincide with food shortage or aggression, and perhaps may be prompted by the need to avoid inbreeding. This would also seem to be a risk for fathers and stay-at-home daughters. However, that risk may be low because several males compete for each female, and fathers often do not live long enough to sire more than one litter by their daughters.

For dispersing youngsters the problems of finding a territory have increased in the modern world as wilderness has dwindled. In addition, human persecution of these massive Carnivores has continued since prehistoric times. In Europe, about a dozen Brown Bears cling on in Italy's Abruzzi mountains, where farmers are fully compensated for any damage they cause. In the Pyrénées two populations total twenty to thirty individuals, and forty to fifty more live in two populations in the Cordillera Cantabrica of northern Spain. Their crumbling foothold on the planet illustrates all too well the precarious situation of all large Carnivores. Nonetheless there are strongholds in eastern Europe, particularly Russia, and the 1991 estimate of a total of 35,000–40,000 European Brown Bears is an increase on the recent past.

In the Far East, the predicament of bears is worsened by the high value placed on the supposed medicinal properties of their gall bladders, and on their *haute cuisine* paws. The tradition of relishing bears' paws stems from the myth that bears survive the winter sustained only by sucking the pads of their feet. This story arose because old, calloused skin on paws may slough off in winter dens leaving tender pads. People reasoned that if pads could sustain a mighty bear over winter, they could greatly invigorate somebody eating them, and paws acquired fantastic value. Japanese importers pay US$75 per kilogram ($34 per pound) for Chinese bear paws, and sell them wholesale for $100 each to Tokyo restaurants that serve the resulting dish for $850 per portion. In South Korea, an Asian Black Bear was auctioned for US$18,500, and an undried bear's gall bladder fetched $55,000. In the late 1980s six bear farms were established in Heilonjiang, China, at which the bear's gall was 'milked' periodically via tubes inserted into their gall bladders. Whether this will be successful, or reduce the wildlife trade, remains to be seen.

In North America, settlers killed thousands of Black Bears for meat, fat and fur, and cleared their forest habitat for farming. Attitudes were jolted in 1902, when President Theodore Roosevelt refused to shoot a Black Bear chained to a tree by toadies determined to ensure that he would not miss. His mercy caught the public's imagination and teddy bears were invented to celebrate his compassion. Today, however, American hunters are licensed to crop thousands of Black Bears and hundreds of Brown Bears annually. Bears also clash with foresters, bee-keepers and picnickers. In early summer, after a winter of fasting, Black Bears strip the bark off prized timber to reach the nutritious sap underneath. One bear on a binge can kill 50 trees (worth US$20,000) in a night. Until recently foresters shot the culprits at a labour cost of about $600 each. Now they are exhibiting an ingenuity that should inspire all Carnivore managers: they feed the bears during the bark-stripping period, which is both effective and costs only $50 per bear.

Problems with picnickers began in the 1930s, when bears feeding on refuse in Yellowstone and Yosemite National Parks became a major attraction. The bears lost their wariness of people, and people lost their fear of bears. The bears came to expect hand-outs and, when these were not forthcoming, raided picnic tables, tore open tents and broke car windows. One Yosemite Black Bear specialised in mugging Volkswagens, which are airtight when their doors are locked and windows rolled up. The enterprising bear would climb on to the car roof and jump up and down until the air pressure burst the doors open, allowing the bear access to the hampers within. So successful was this bear that it took to bouncing on every VW it could find … until it was shot.

After bitter controversy park refuse dumps were fenced in 1971 and hundreds of bears starved, whilst others driven to search for food outside the parks were shot. In 1972, there was the first fatal mauling of a person by a bear in 30 years. Today the problem continues, with Brown Bears mauling about six people a year in the USA. Paradoxically, Black Bears cause more injuries because they are less aggressive and therefore treated more casually. Nuisance bears are sometimes moved, often to little effect: 12 Alaskan Brown Bears transported over 200 kilometres (125 miles) returned home in under two months. Another solution being tried is training. At a 'bear school' in Montana, for example, each new problem bear is introduced to a tutor, who immediately sprays red pepper oil (capsaicin) in its face. The pepper is irritating, but not dangerous, and after a few tutorials even the most recalcitrant bears get the idea and run at the sight of people. So far, over 90 per cent of problem bears have been released as reformed characters. In Denali National Park, Alaska, problem bears are fitted with radio-collars so that a trainer can find them. Every so often, the trainer sets up a tent nearby and, if one ventures close, strafes the intruder with plastic bullets. These bears soon give up raiding camps.

Thanks to its icy home the Polar Bear has a sanctuary largely out of human reach, although aeroplanes and snowmobiles are allowing hunters increasing access. The Polar Bear is one of the most recent species of Carnivore, having arisen only 250,000 years ago from a northern population of Brown Bears that gave up omnivory to concentrate on seals. It was a small step, because Grizzlies in northern Canada often eat seal carcasses on the sea ice, and occasionally hunt seal pups. The two species can still interbreed and female offspring of their liaisons are certainly fertile, although males may not be.

The southern limit to the Polar Bear's range has varied with the comings and goings of 17 Ice Age glaciations. About 70,000 years ago, they were roaming the frozen wastes south of London. Their northern limit lies at 82 degrees, which is the boundary between the shallow continental shelf waters, where North Atlantic

(Overleaf) American Black Bears have adapted to the abundant food refuse in the vicinity of camp grounds, but this habit has brought them into perilous contact with people

currents ensure high biological productivity, and the relatively lifeless Arctic Ocean. There the ground is often, literally, shifting under the Polar Bear's feet. Cracks in the sea ice, where seals can be caught, constantly appear and disappear. On a longer time-scale, an area where seals are abundant one year may be devoid of them the next. Like other temperate bears, the Polar Bear faces protracted lean periods, but unlike its onnivorous relatives its leanest period is in late summer, when seals have dispersed.

The largest recorded Polar Bear just outweighs the largest Kodiak Bear at 1002 kilograms (2204 pounds). However, adult male Polar Bears generally weigh 350–650 kilograms (770–1430 pounds) and females 150–250 kilograms (330–550 pounds).

The transformation of Brown Bears to Polar Bears involved many adaptations to an icy and carnivorous lifestyle. They grew creamy white fur everywhere excepting the footpads and nose. They lose heat from these sites when hot and curl up with a paw over their muzzles when cold. The shift from fruit and nuts to flesh led to the abandonment of grinding cheek teeth in favour of smaller, jagged scissor teeth and large canines in sharper muzzles. Polar Bears have shorter, more solid claws, less likely to break on ice, and huge feet that can act as paddles in water and spread like snow-shoes on thin ice. Their footpads are pimpled like the surface of a table-tennis bat to grip the ice, and little depressions in the sole of the foot may add traction. On very thin ice, Polar Bears crawl on their knees and elbows to spread their weight.

Polar Bears' bodies are covered with a layer of blubber up to 11 centimetres (4.3 inches) thick which, combined with their amazing fur and black skin, provides such excellent insulation that they feel cold in only the most cutting wind. They appear black on ultra-violet (UV) photos, because their fur absorbs UV light so well. Each hair is hollow and acts like a fibre-optic filament, conducting the warming UV light to the heat-absorbing black skin. Short-wave UV radiation passes through clouds, so the bears can warm up even on overcast days. On a sunny day, a bear's skin may be warmer than deep within its body.

With such marvellous insulation, Polar Bears are in more danger of overheating than chilling. This is why they plod around so ponderously, and large males and pregnant females in particular rarely gallop for more than one or two minutes at a time. Large males overheat so quickly that Eskimos on foot can run them down in a few hours. The risk of heat stress may explain why large males have large roman noses. The snout radiates heat like a beacon. It probably also acts as a heat exchanger to retain heat, containing convoluted bones coated in expansive moist membrane. As the bear breathes in, chilled dry air is warmed and moistened before it reaches the delicate lungs, and as the animal breathes out, the heat and water are recaptured again by the membranes. Another problem arises from their stocky body and massive limbs and paws, which necessitate a lurching gait that consumes twice the energy most animals use when walking. Thus they wait for prey rather than chasing them. The heat of the chase is so great that they cannot afford to take flightless Snow Geese in summer: a chase of 12 seconds would consume all the energy the bear would gain from eating a goose.

Polar Bears feed almost entirely on seals ambushed as they surface to breathe at cracks in the ice. With one blow, the powerful bear can scoop a seal out of the water and kill it. Their vision is similar to ours, but they can sniff out seals 1 kilometre (0.6 miles) away under 1 metre (3.3 feet) of snow. Days may pass between kills, but Polar Bears compensate by gorging, sometimes on 15–20 per cent of their weight in one meal. They are particularly efficient at digesting protein and fat, and during the spring glut of seal pups they may eat only the fat. With stunning powers of navigation on the shifting sea ice they can trek 30 kilometres (19 miles) a day seeking the seals' favourite sites, which may shift by more than 100 kilometres (60 miles) between years. In areas such as the Barents, Greenland, Chukchi and Bering Seas, Polar Bears roam thousands of kilometres each year in search of seals.

Apart from a long association between mother and cubs, Polar Bears are solitary. Their year begins in February or March when mothers and cubs emerge from winter dens and go on to the sea ice. In late March and early April, tens of thousands of female Ringed Seals dig birth lairs in snowdrifts over breathing holes in the ice and there give birth to pups. Polar Bears gorge on the corpulent pups, and rapidly put on weight to sustain them over the lean summer months when the seals disperse to open water.

Male and female bears meet on the sea ice, wherever seal hunting is best at the time. Overall, their sex ratio is 1:1, but because the females nurse their pups for 2.5 years only a third of the adult females are available each year. Between April and May, a male will scour the ice, seeking out the tracks of females and identifying, presumably by their perfume, those that are receptive. When he eventually catches up, the male may find up to six other suitors gathered around to fight it out. The resulting competition explains why males are two to three times heavier than females, and heavily battle-scarred. The winner is likely to be at least eight to ten years old. He leads the female away from the prime feeding grounds and thus from the bombardment of other males that congregate there. The pair remains together for a week or more, mating many times a day.

As the ice begins to retreat northwards, from late May to July, more northerly Polar Bears follow over several hundred kilometres, swimming up to 40 kilometres (25 miles) between ice floes. More southerly bears, and all pregnant females, cannot follow the ice and must live for the rest of the year on fat stored during spring. An adult Ringed Seal is the weight of a person and normally feeds a large bear for about 11 days, but while fattening up a bear may eat one every four or five days. A female anticipating pregnancy will gain 200 kilograms (440 pounds), doubling or sometimes quadrupling her early spring weight while food is plentiful. When the males and non-pregnant females leave, the pregnant females den up in snowdrifts for an eight-month confinement. In autumn the northerly contingent of bears travels south again in the wake of ice and seals. As the sea freezes over they again hunt on the sea ice while the pregnant females sleep on. The temperature inside the maternity den falls below freezing but, as in an igloo, remains about 20°C (36°F) warmer than that outside.

The mother gives birth, generally to two cubs, between late November and early January. Each newborn weighs about 0.6 kilogram (1.4 pounds) and measures 30 centimetres (12 inches) long. Drawing on her fat, the mother nurses the cubs until spring, when each is between 10 and 15 kilograms (22–33 pounds) and the size of a small dog. The milk is uncommonly rich, containing almost 50 per cent fat against 11 per cent for a Sun Bear's. This allows the cubs to balloon from rat-size to person-weight in one year. By the time they emerge from the den, male cubs already weigh about 10 per cent more than their sisters. Within weeks, both begin the long trek to the sea ice and their survival hangs in a balance weighed by seals. Where seal pups are abundant, bear cub mortality is low, and vice versa. Those that survive adolescence may live 30 years.

Pregnant females in dens have a different blood chemistry to active bears. The amount of urea in the blood rises when the animal is eating and becomes very low when it is living on its own fat. Another substance in the blood, creatine, stays constant. So the ratio of creatine to urea in the blood shows how active or lethargic a bear is. Denning female Polar Bears have a fasting or 'sleepy ratio'. The southern Polar Bears that do not retreat to dens but must fast through the late summer also have a sleepy ratio at that time – biochemically, they are asleep on their feet! Denning bears are also metabolic marvels. Their kidneys run slowly for months at a rate that would kill people, and they maintain their skeletal strength over months of lassitude while the bones of bedridden humans become brittle.

A similar state of lethargy occurs among bears that rely on fruit and nuts, and live in areas where such foods are scarce in winter. These bears, too, usually den alone. European Brown Bears feast in autumn on beech mast and acorns, supplemented by sweet chestnuts and hazelnuts. They then retire to caves or hollow trees for three to six months, depending on latitude. Black Bears in the harsh conditions of northern Minnesota den for five to seven months of the year, while their pampered southern cousins of coastal Washington and North Carolina den for less than half as long, and further south bears do not den at all. Denning Black Bears may devote 20 hours a day in autumn to stocking up on acorns, beechnuts, berries, apples, insects, carrion and whatever else they can find. In the north, where fruit is less abundant, populations have to eat bulk supplies of leaves and shoots to compensate, but have difficulty in processing sufficient of this less digestible food. Each adult consumes about 20,000 calories a day, five times its normal intake, building up a 13-centimetre (5-inch) thick layer of fat around the body. The bear retreats into a den when its daily intake of calories falls below the effort expended in foraging, so that it would be better off doing nothing.

In hibernators, such as Hedgehogs, body temperature can drop to a few degrees above freezing, and the metabolic rate plummets. Carnivores do not truly hibernate, although the difference seems to be only a matter of degree. Their body temperature only drops from 38°C (100°F) to 31–35°C (88–95°F), while their metabolic rate idles at 50–60 per cent of the normal. A torpid Black Bear can usually arouse fully in only a few moments. One biologist leaning his head against a dormant Black Bear's chest was alarmed to discover that it only takes a few seconds for the heart rate to accelerate from a soft eight beats a minute to a thumping 175 beats a minute.

Like the Polar Bear, female Black and Brown Bears give birth during denning, the mother only rousing momentarily for the task. Newborn cubs are tiny and almost naked, and nursing them is a huge drain on the mother's energy and water resources. A mother Black Bear emerging from the den with three-month-old cubs will have lost up to one third of her weight. Northern bears must be born early in the new year to be old enough to take advantage of the springtime flush of food and lay down sufficient fat reserves to survive their first winter. The bear's gestation takes three months, so meeting the birthday deadline would require mating in October when males and females are asleep in separate lairs. Smaller berry-eaters, like raccoons, can squeeze a pregnancy in quickly in the spring but bears have had to find another solution to the incompatibility of sex and sleep. Brown Bears mate in May or June, and the fertilised ova begin development. However, the embryo soon falls into a state of suspended animation until some five months later, well into the mother's winter lethargy. Then it implants in the uterine wall and continues its development normally. This is called delayed implantation. The tiny embryo imposes minimal burden on its mother until it is reactivated to keep its appointment with the world. If the mother fails to eat enough during the autumn because the berry crop is insufficient, the embryos will not implant.

Although the procyonids have a similar diet, they differ from the bears in many ways, including size. Many procyonids do not need to retreat in winter because, like the Olingos and Kinkajous, they live in the tropics where fruit is available all year round. The northerly bears are not only forced into a winter sleep, but also, since lethargic Carnivores have higher temperatures than true hibernators, they face a big problem of heat loss. This favours a large body, which loses heat more slowly. Yet other creatures that face the same problems have not grown as massive as the bears, possibly because they need to gather other foods, climb trees or enter burrows. Another way to conserve heat is to cuddle up together, and this is what raccoons do. As many as 23 raccoons have been found in one nest, usually including only one adult male. Similarly Eurasian Badgers, which eat a lot of plant material by the standards of the weasel family, den communally over winter, as do the berry-eating Asiatic Raccoon Dogs.

While the need to conserve heat will have increased the advantage of bears growing big, it is not sufficient explanation for their large size. The Florida Cave Bear lived in warm climes yet was as gigantic as its European counterpart in the cold. The omnivorous bears were able to grow big because their diet of fruits, tubers and occasional flesh was sufficiently abundant to support big appetites. Indeed, temperate plants produce such an abundance of autumnal fruits and nuts that a determined bear can stoke up enough food to run its body for months. While great size was no impediment to their foraging it was useful in securing food and fighting off competitors. Furthermore, the larger bears grew, the fewer predators they had.

Large size has disadvantages too, including a slow reproductive life and the need for a huge range, and these factors have also affected bears' social lives. Each female bear has a large territory in which she cohabits continuously with successive generations of adolescents. Since she has outgrown most predators and feeds mainly on vegetation, there is little the male could do to help her, even if he could share her territory alongside her daughters. Instead, he tries to mate with as many females as possible, which has put a premium on fighting attributes, such as immense size. Since the territories of, say, three breeding females, would be undefendably large, he wanders promiscuously.

There may be many variations on this general pattern because few bear societies have been studied in any detail. For example, it is possible that the termite-eating Sloth Bear has a social life more like that of the termite-eating Aardwolf (see page 123). The more similar sizes of male and female Sun Bears may indicate that they are more monogamous. The exaggerated size difference between the sexes of Spectacled Bears may exist because their food supply fosters extreme polygyny or may be partly an heirloom from their carnivorous running bear ancestors. Certainly, bear society, like that of other Carnivores, is flexible. Despite their customary stand-offishness bears, like Raccoons and Kinkajous, will congregate at rich pickings. What matters is the pattern in which food is available, rather than whether it is fruiting trees, picnic hampers, leaping salmon or pupping seals.

Two Carnivores, the pandas, have escaped the seasonal vicissitudes of temperate fruits and nuts by eating evergreen foliage. The Red Panda's flat and broad cheek teeth, with their complex pattern of ridges, would look more at home in an antelope's mouth than a Carnivore's. In its native forest habitats in Nepal, Burma and China, the Red Panda feeds largely on bamboo, supplemented with fruit, roots, acorns, lichens, and the occasional mammal. Little else is known of its habits because it is scarce in the wild, and may be rare even in its strongholds, with perhaps only 24 surviving in Nepal's Lantang Reserve. However, 500 exist in zoos, where captive breeding is succeeding.

The Giant Panda, once widespread in mountainous China, is now facing extinction with less than 1000 individuals surviving in the bamboo forests of Szechuan,

The Red Panda is linked to the Giant Panda only by its English name and adaptations to eating bamboo. Its lineage stems from the roots of the Procyonid family

Shensi and Kansu provinces. Apart from an occasional lapse, for example stamping outside a Bamboo Rat's nests to make the occupants bolt into its jaws, the Giant Panda is also vegetarian. It concentrates on bamboo and has an oesophagus with a tough lining against splinters, and a stomach with a muscular construction like a gizzard for pulping bamboo. Despite these adaptations, it is still inefficient at digesting bamboo, which is a legacy of its carnivorous past. Bamboo and other plants are made up of cells that are a bit like hot-water bottles, with fluid contents and a tough wall. While some of the goodness in leaves slops around in the juices within each cell, it is mainly locked up in the walls. The Giant Panda can digest about 90 per cent of the nourishment inside bamboo cells, but only 17 per cent in the walls. In contrast, animals from a long line of vegetarians, like antelope, can digest 90 per cent of the walls.

Giant Pandas compensate for their poor digestion by bulk loading, dedicating their waking hours to seeking, picking, and eating bamboo. They shift food fast through the gut, passing bamboo shoots through in five hours. This dedication and speed allows an averaged-sized, 100-kilogram (220 pound) Panda to consume almost 8 kilograms (18 pounds) of bamboo daily. However, it extracts from the bamboo little more energy than that needed to obtain and process the food. So, Giant Pandas live precariously close to energetic bankruptcy and conserve energy by sleeping whenever they are not eating. They cannot afford to hibernate, instead moving to lower altitudes in winter.

In addition to their natural problems, Giant Pandas suffered in the past from hunting. A notorious case involved Theodore Roosevelt's sons, Theodore and Kermit, who tarnished the family's 'teddy-bear' image by becoming the first westerners to shoot a Giant Panda. An even graver threat has been the loss of bamboo to agriculture. This has affected Giant Pandas particularly seriously, partly because of a quirk in the life cycle of bamboos. There are seven species of bamboo, each growing at a different altitude. Normally they all grow abundantly and provide a reliable supply of food. Every 40–120 years, however, all the bamboo plants of a certain species in a certain area will flower and die. It takes 10–15 years before the new shoots have grown sufficiently to support Giant Pandas. Previously there were huge tracts of bamboo forest and Giant Pandas could decamp when one area flowered. Now, Pandas are isolated in six small enclaves along the eastern rim of the Tibetan plateau, each a cluster of mountain tops containing only one or two species of bamboo. The isolated populations often number fewer than 50 individuals and there are no corridors between them. Worse, in the wake of deforestation local hunters set snares for Musk Deer, and these snares became the biggest cause of Giant Panda deaths. By the mid-1970s fewer than 2000 survived.

Then the Chinese imposed the death penalty on anyone convicted of killing Giant Pandas or smuggling their skins. Three people have been executed to date, but poaching continues at a dangerously high rate because of the huge financial incentive. By 1978 there were 12 panda reserves and a research centre at the 2070-square-kilometre (800-square-mile) Wolong Natural Reserve, which currently supports over 100 Giant Pandas. One aim at Wolong is to relocate the mountain farmers to free lowland accommodation but, not surprisingly, the villagers are reluctant to move. Another aim is to restore in the wild the seven species of bamboo currently being grown in nurseries, so that each population of Giant Pandas will have access to at least two species of bamboo. The Chinese conservation plan formulated in 1991 called for 14 more reserves with corridors between them to add 1900 square kilometres (700 square miles) of panda habitat to the 3350 square kilometres (1300 square miles) already protected.

The fragility of the wild populations has made captive breeding a vital backup, but this has proved difficult. In the wild, Pandas are slow breeders, raising at most one cub every two years. After delayed implantation and a pregnancy of 3.5–5.5 months, young weighing 90–140 grams (3–5 ounces) are born almost naked in a cave or hollow tree. Although twins are common, a mother never attempts to rear more than one, and for three weeks she cradles the baby, sitting upright, almost continuously. It is 2.5 months before the cub can stand and 18 months before it leaves the female. In captivity, breeding is even slower. The first problem is that pandas are difficult to sex because the male's penis is normally hidden and he has no scrotum, the testes remaining inside the body, surrounded by fat. Then, the female is receptive for only a few hours annually, and is fussy about her mate. Artificial insemination has been thwarted by difficulty in detecting the female's fleeting heat and only three litters had been produced in this way outside China by 1989. Captive mothers often desert their young and so far all hand-reared newborns have died within 45 days. Apparently unless gripped firmly and constantly, the cubs wriggle and fret until they die from exhaustion. In addition, there are few substitutes for Giant Panda milk, which is oddly low in sugar. As a result of all these difficulties, it took from 1937 until 1963 for the first successful captive rearing, in Beijing Zoo. Today there are only 110 Giant Pandas in zoos (almost 100 in China), and they are dying faster than they breed.

Captive Giant Pandas have been better at raising money than young. When Tokyo's Ueno Zoo had an artificially inseminated Giant Panda birth in 1986, over 270,000 people suggested names for the cub (the winning name being Tong Tong which means 'child'!). Nearly 13,000 people queued to get a glimpse at the first public viewing of the cub, and 200,000 called the 'Dial-a-Panda' hotline *each day* to hear its squeals. As long ago as 1958, Chi Chi netted US$2000 a week on her tour of Europe, which put her value at four times that of the then American secretary of state. At 1989 prices, a 3–6-month loan by China of a pair of pandas would cost US$300,000–600,000. In return, a pair visiting San Diego zoo in 1987–88 boosted attendance by 35 per cent and increased income by over US$5 million.

The arrival of this enigmatic creature's cousin, the Red Panda, in London began a debate about where the pandas fit into the Carnivores' story. Most palaeontologists thought Giant Pandas were bears, while anatomists argued that they were weird procyonids. Eventually Red and Giant Pandas were lumped in a family of their own, the Ailuropodidae, although the evidence of their kinship was weak. Then, after 120 years of contention, the vexed question seemed to have been resolved by molecular techniques, the biological detective's equivalent of forensic evidence. In 1985 biologists compared proteins from both pandas with those of Brown, Spectacled and Asiatic Black Bears and Common Raccoons. They concluded that the Red Panda split off from the procyonids 28 million years ago, while the Giant Panda split off from the early bears about 22 million years ago. So the Red Panda was a raccoon, the Giant Panda a bear, and their similarities lay only in their superficial appearance and English name. Scientists had scarcely digested these results when, one month later, another study revealed that the haemoglobin in the blood of Red and Giant Pandas was strikingly similar, and different to those of bears and raccoons. This may be due to the fact that both species of panda are adapted to living at very high altitudes, just as the similarity in their 'thumbs' may be independent adaptations to grasping bamboo. Perhaps the thumbs first evolved as climbing irons since both species are tree-climbers. In 1989 a further bumper analysis of 289 proteins confirmed that Red Pandas are early procyonids, whereas Giant Pandas are more recent bears. However, the debate is unlikely to be over.

Previously the Asian Black Bear had been considered the most vegetarian bear, while the nectar-eating Kinkajou had been the least carnivorous procyonid. Now these titles are usurped and the two pandas stand as monuments to Carnivore adaptability. Leaf-eating antelopes and monkeys generally live in groups, there being little disadvantage to feeding side by side, and great advantage to shared vigilance. The Giant Panda does not, possibly because it has outgrown the need for vigilance. Female Giant Pandas browse alone in 30-hectare (75-acre) enclaves within shared 4–6 square kilometre (1.5–2.5 square mile) ranges. Males are about 10–20 per cent bigger and seem to have a pecking order to settle disputes, for example when several congregate around a female on heat. The system is uncannily like that of the European Elk, another huge browser that has outgrown most predators. Female Elks munch alone, visited by males that are 25 per cent larger and compete in shoving matches.

Because of its eccentricities, articles frequently refer to the Giant Panda as an over-specialised relic, incapable of adapting to the modern world. Both pandas do fall outside the Carnivore norm in being born light and growing slowly, probably because bamboo is low in protein. However, Giant Pandas are no more anachronistic than any other Carnivore. Allusions to their 'failure' in the modern world probably arise out of people trying to convince themselves that the Giant Panda's imminent extinction is not our fault. But, like the plight of the other large bears that opted for life in the slow lane, it is our fault.

Family Tree: Bears and Procyonids

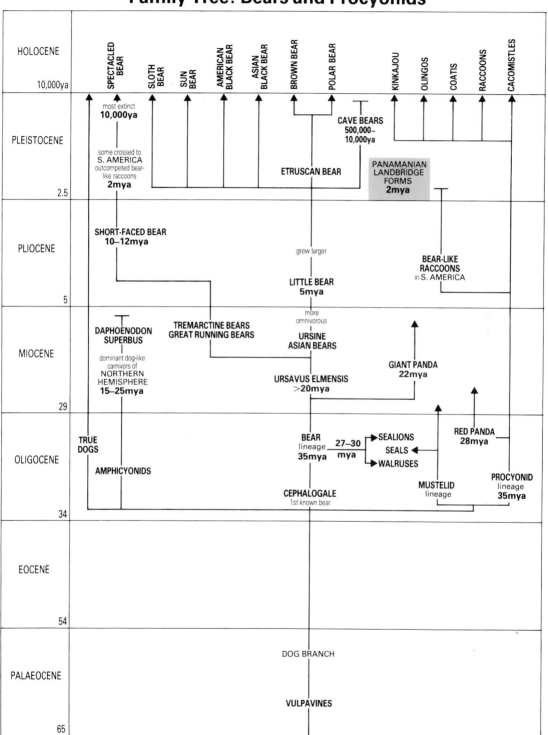

CHAPTER 6

SMALL AND DEADLY

The Mustelidae is often referred to as the weasel family, but in addition to weasels, polecats and martens, it includes badgers, skunks and otters. The family is named in deference to the fragrances emanating from its members' anal glands. Although all are rather small, mustelids are the most diverse Carnivore family. The heaviest, the 30-kilogram (66-pound) Sea Otter, is 1000 times heavier than the Least Weasel, the smallest Carnivore ever. Mustelids also differ enormously in the size of prey they take relative to their own size. A Stoat or Ermine can catch rabbits weighing up to 10 times its weight whereas a Eurasian Badger may live on earthworms 1/3000th its weight, giving a 30,000-fold range in relative size of predator and prey. The largest prey taken by mustelids are probably the 272-kilogram (600-pound) Caribou caught by Wolverines. The mustelids' diversity owes much to their long thin bodies which have been adapted by different branches of the family to allow them to tackle prey in trees, underwater, on the ground, through crevices and down burrows.

Mustelids are the most recent offshoot of the dog branch, sharing an ancestor with the procyonids, and are distinguished from the other dog families by having one upper and two lower molar teeth in each side of the jaw. Although the dog branch began in North America, much of the early diversification of mustelids, along with that of the related procyonids and bears, was in Eurasia. The shift to a cooler, more temperate world climate about 30 million years ago accelerated the opening of forest glades and savannas where small herbivorous mammals prospered. These prey could escape underground from early terrestrial Carnivores such as *Cephalogale*. This wolfhound-sized early bear could only wedge its nose in the burrow's entrance and drool over the appetising odours within. But a contemporary, *Palaeogale*, was at the base of a lineage that would solve this problem. *Palaeogale* was marten-sized and had the sinuous climbing build of the earliest Carnivores. Unearthing prey was to become one of the specialities of its descendants.

An early strand of mustelid evolution opted for the brute strength of professional

Pine Martens have much in common with the arboreal ancestors of the mustelid family

diggers. About 22 million years ago, *Aelurocyon brevifacies*, the biggest ever mustelid, had powerful jaws and a thick-set demeanour like its cousins, the bears. The size of a Puma, with broad forepaws and raking claws, it was reminiscent of a lineage of diggers that, seven million years ago, gave rise to *Pliotaxidea*, ancestor of today's American Badger. *Pliotaxidea* was long-bodied and massively muscular, with broad, flat feet and alarmingly curvaceous claws. American Badgers weigh between 3.5 and 12 kilograms (7.7 and 26 pounds), and are greyish to reddish above and buff below, with dark brown feet. Their most striking feature is a white stripe from nose to shoulders. Professional diggers with compact bodies, their stout forelimbs and loose skin aid turning in tight corners. Adults feed mainly on ground squirrels but youngsters lack digging skills and so feed more on insects. When winter snow makes prey unavailable badgers doze until the weather warms. Outside the mating season males' home ranges are about 170 hectares (420 acres) and slightly bigger than females'. Females tend to be territorial, but males' ranges overlap each other and those of females. In the rut males roam far beyond their normal ranges, mating with several females, and neighbourly relations are so strained that many, especially males, bear battle scars.

Despite its name and lifestyle, the American Badger may not be a true badger. Most badgers occur in the Old World, where eight species of them are grouped in a subfamily known as the Melinae or honey-eaters. All these true badgers are medium-sized, thick-necked, stocky animals, with powerful jaws containing broad, crunching cheek teeth. They have long rootling snouts, taken to an extreme in the mobile nose of the Hog Badger from Asia. With the exception of three species of ferret badger from tropical Asia, all have short tails. Their forepaws are animated spades, with long, non-retractile claws. The toes of the trowel-like forefeet of the Malayan Stink-badger are united up to the claws.

Badger evolution involved the progressive abandonment of the carnassial scissors in favour of broad back-teeth suited to grinding vegetable foods. This culminated in Eurasian Badgers, the least carnivorous mustelid, which probably originated in China about two million years ago. Its jaws clench shut so powerfully that the lower jaw is locked by a bony hinge to the skull to prevent its disarticulation. This 12-kilogram (26-pound) muscle-bound omnivore eats berries, tubers, small mammals and insects but where possible specialises on earthworms. Eurasian Badgers have been reported in pairs in Spain, solitarily in Apennine Italy, and in parts of Sweden the territory of one male may overlap several female territories. However, over much of Europe they live in mixed-sex clans of about 10 members (maximum 25). Members of a clan cohabit in a large den, or sett, within a shared territory ringed by dung-pits. Dominant males daub scent from pockets below their tails at these latrines and on the bodies of other clan members. Bacteria in the pockets perfume the scent, and occasionally clan members press their backsides together in a kiss that leads all clan members to share the same bacteria, and,

One early lineage of the mustelids, known as the oligobunines, started as rather heavily built creatures. Perhaps their ancestry stemmed from creatures like this Cephalogale, an ancient bear. For a creature like this there were to be two evolutionary routes ahead when their prey escaped underground: one was to become professional diggers, the other to become burrow-hunters

One lineage of early mustelids evolved to become thick-set, brawny diggers with raking claws. Aelurocyon *(far left)* was in this mould, as are modern badgers and, to an extent, Wolverine and Ratel. Another approach was to evolve towards tubular bodies that could squeeze into the prey's sanctuary, and this was the route taken by Zodiolestes *(left and below)*

consequently, the same odorous membership badge.

Eurasian Badgers can only feed on worms when the ground temperature and humidity are right for them to surface. The best worming places vary from night to night, depending on the weather, so a badger must defend a territory large enough to ensure that somewhere within there will always be worms available. However, when worms surface they often do so in thousands, so plenty of badgers can share the spoils. The result is that the smallest territory that will satisfy one badger may be enough to sustain a large clan. In these stable clans, males and females associate for many years, but their apparent faithfulness may be deceptive. Blood tests have shown that some female Badgers bear litters sired by two males. Furthermore, in a substantial minority of clans, some cubs could not have been fathered by the resident males, whereas an incriminatingly appropriate father lived in the neighbouring clan. It seems that illicit midnight trysts are commonplace.

Following an autumnal feast badgers spend the winter lethargically in their setts. Some are immense: 180 entrances, 880 metres (960 yards) of tunnels and 50 nest chambers were recorded in one. Nest chambers contain room for only two or three adults and are stuffed with grassy bedding dragged in by the badgers (along with 250 golf balls in one sett!). In cold weather badgers share nest chambers, the decaying grass probably providing central heating. Some tunnels are stuffed with bedding, perhaps as a store or perhaps as a further source of heat. Members of a clan may have favoured sleeping partners, but all move nest chamber often, perhaps to limit the build-up of fleas that throng the burrows. The intricacies of temperature control and flea-avoidance may explain the size and complexity of setts. Badgers cannot settle where the ground is unsuitable for sett-digging, and once a sett has been dug it represents a huge asset and may be inhabited for centuries.

While badgers mastered omnivory, another subfamily, the Lutrinae or otters, adapted the tubular shape of their ancestors to become aquatic hunters. The first mustelid to test the water was probably like today's mink which are members of the weasel subfamily. Modern mink usually live in river banks and hunt aquatic prey but are at least as at home on land. This compromise means, for example, that mink have eyes suited to terrestrial rather than aquatic vision. Modern otters have almost three times better visual acuity underwater than mink. The Asian Clawless Otter and Sea Otter can potentially see equally well in air and water because their enormously developed iris muscles can squash the lens in each eye into a spherical shape suitable for underwater vision. In addition, otters have huge whiskers on muzzle and elbows to feel their way in murky water and to detect turbulence.

The otter lineage also gradually evolved lithe bodies with flexible spines for rapid manoeuvrability, and muscular tails and webbed feet for thrust. Otters are

(Left) River Otters and their lineage are closer to the badger lineage than to the other mustelids

(Overleaf) Eurasian Badgers are unusual amongst mustelids in living in social groups

streamlined, their necks wider than their heads, and they have dense underfur and long water-repellent guard hairs. The earliest fossil otter, *Potamotherium*, lived in the Loire Valley 25 million years ago, and was probably the ancestor of seals (see page 158). The earliest ancestor of modern otters might be *Mionictis*, whose fossils date to 17 million years ago in North America. Today, the most aquatic species is the Sea Otter, which has such large hind flippers that it is cumbersome on land, and a flattened tail which acts as a rudder. River Otters swim using only their hind limbs, but Sea Otters also use vertical movements of the tail.

As ancestral otters mastered the water some, like the Spot-necked Otter, became streamlined torpedo-like predators of fish. They detect prey using their keen eyesight. Others, like the Cape Clawless Otter, took to raking the mud for shellfish. Spot-necked and Cape Clawless Otters both live in African lakes, but the mud-raker has much longer whiskers and more dexterous fingers. Its fingers lack webbing and have sensitive tips to sift and probe the sediment, and its claws are like human fingernails.

The teeth of torpedo and mud-raking otters also differ. The fish-eaters have sharp incisors and canines to hold prey, and shearing scissor teeth to cut it up. They often carry prey in the mouth back to land, where they hold it down with one forepaw while tearing pieces off with their jaws. The flesh of fish is easily digested but the bones are dangerous. Giant Otters of tropical American rivers eat fish head first and produce copious mucus to protect their guts, through which the flesh can pass in only half an hour. The invertebrate-eaters grasp shellfish in their sensitive hands and pass them to the back of the mouth where there are flattened crushing molars designed to splinter shells. They rarely carry food back to land but when they do so they carry it by hand.

The Sea Otter, which ranges around coasts from California around Alaska to Kamchatka and Kuril island, has developed another technique for coping with its diet of sea urchins, clams and abalone. It uses stones to smash clams, and keeps a favourite stone in a 'waistcoat pocket' – a fold of skin under its armpit. It may also store food in the pocket, to be retrieved when it needs a snack. Newly established Sea Otter populations specialise on favoured prey such as sea urchins and abalone, but once these are reduced the otters must turn to a more generalised diet. However, in one long-established population, in Monterey Bay, California, individual female Sea Otters specialise in types of prey. Even foraging side by side these females have different diets. Some concentrate on prey like turban snails, a pocketful or dozen of which can be gathered in just over a minute, but then each requires six whacks or more with an anvil to smash it. On average it takes two minutes to smash the whole catch on the surface. Others favour prey that are harder to find, like abalone which take up to 63 dives totalling 30 minutes and underwater pounding with a rock to dislodge, but are then easily eaten and digested. Daughters may adopt their mother's speciality, perhaps through copying them. Consequently, although the average Sea

Otter at Monterey appears to be a generalist eating 27 prey types, individuals have a much smaller range of prey, a maximum of nine types, one of which may comprise as much as 81 per cent of the diet.

Most otters make short dives, the averages for River Otters (15 seconds) and Giant Otters (25 seconds) being not much greater than that for the American Mink (10 seconds). A Sea Otter's lungs are three times larger, relatively, than those of either a River Otter or seals. Yet the Weddell Seal commonly dives for 30 minutes (maximum 73 minutes), while Sea Otters generally dive for under 2 minutes to about 20 metres (65 feet). So perhaps their capacious lungs are used more for buoyancy than for breathing. Baby Sea Otters, which cannot swim until 6–8 weeks old, are so buoyant that they cannot sink. Sea Otters are so fully aquatic that they can survive without ever hauling out on land.

Heat is lost to water 25 times faster than to air and smaller animals lose heat faster than larger ones. Otters, therefore, cannot afford to be small. Their average weight is about 13 kilograms (29 pounds), while that of other mustelid species is less than 3 kilograms (6.6 pounds). Furthermore, turbulent water carries off heat up to 100 times faster than air. So the Sea Otter, which spends its life at sea, is the heaviest, being up to 45 kilograms (100 pounds) in weight. It is a puzzle how the 4-kilogram (8.8-pound) Chilean Marine or Feline Otter, which also spends much time at sea, gets away with being the size of a new-born Sea Otter pup. Sea Otters and fur seals are the only marine mammals to rely on fur rather than blubber for warmth. Sea Otters have the densest hair of any mammal with, on average, 126,000 hairs per square centimetre (813,000 per square inch), twice as dense as that of fur seals. Sea Otters devote three hours daily to blowing into their coats to fill them with insulating air, and have especially loose skin to allow them to reach all parts of the body with the nose. Their fur helps keep them afloat and resists water, maintaining a warm-air blanket around them. As a result, they lose heat in water only twice as fast as in air, and at a tenth the rate of normal mammals in water. They make up for the loss by producing more heat through a souped-up metabolism that runs at 2.5 to 3.2 times the rate expected for a similar terrestrial mammal. This internal fire needs fuel and the Sea Otter eats voraciously, consuming 23–33 per cent of its own weight daily. To reduce the punishing heat loss to the ocean, baby Sea Otters need to grow very fast, and so are fed milk exceptionally rich in fats.

Otters' feeding habits have shaped their societies as well as their bodies. Many are monogamous, such as Feline Otters which have been seen hunting cooperatively in pairs. But some, such as Eurasian Otters in Scandinavia, are polygynous with lone males occupying large territories that encompass several of the lone females' smaller territories. However, Eurasian Otters living along the coasts of the Scottish Shetland islands live in clans due to their special circumstances. They hunt by day, ambushing fish as they doze in the sea-floor forest of seaweed. The Shetland coast is divided between sheltered and exposed waters. From the former the Otters harvest eelpout, saithe and pollack, and from the latter rockling and sea scorpions. Eelpout are most available in summer and rockling in winter. During November to January, the otters switch from rockling to saithe and pollack, because although the latter are not especially numerous in midwinter, this is the only time they retreat from open water to seaweed forests where the otters can sneak up on them. The variation in prey from place to place and season to season means that each otter must have access to a reasonable stretch of coastline. However, the smallest stretch offering the requisite blend of exposed and sheltered hunting grounds often provides more good fishing sites than one otter needs. Thus two to five females cohabit in a clan territory, while males' ranges overlap each other and at least two female clans.

Females within a clan gang up to repel neighbours, but each fishes alone in her own core area using worn patches in the carpet of seaweed to gain access to the fish's refuges. These patches can be used by only one otter at a time and are fished-out in a few hours and not replenished for about 24 hours. So clan members must space out to avoid visiting a patch that has already been harvested by another. Within a territory, clan size may depend partly on the number of fishing sites but is probably limited more by the number of on-shore peat holes that provide fresh water. Unless Shetland Otters wash regularly in fresh water, their oil glands clog with salt and their fur loses its woolly-vest effect. If the peat holes are used by too many otters for washing, their waters become saturated with salt.

The Brazilian Giant Otter, which is the longest but not the heaviest mustelid, is more conspicuously sociable. Loosely territorial groups of up to six adults fish close together along creeks in tropical forests. Each group seems to comprise a pair and their adult offspring. They declare their territorial ownership by clearing patches of bankside vegetation at sites where they make scoops in the mud and then deposit droppings coated in anal jelly. These, along with urine, form a perfumed quagmire in which the whole family bathes, rubbing the scent-laden mud on each other's bodies. Such sites are clustered in the core of the territory, and the outlying fringes are marked with single faeces. Each Giant Otter has a unique pattern of neck marks, which may help them recognise each other. The 32-kilogram (70-pound) male is scarcely heavier than the female, and both seem to be diligent parents. Certainly both members of a pair joined in killing a keeper at São Paolo Zoo in Brazil whom they apparently thought was attacking their cubs.

The most gregarious of all otters is the Sea Otter. Sea Otters of the North Pacific are divided into northern (Alaskan) and southern (Californian) subspecies. In California the sexes live rather segregated lives. Females occur near the centre of their range throughout the year, gathering in rafts of between four and forty animals, which may be kin since mother, daughter and grand-daughter have been recorded in company. In winter males gather in herds of 20 to 150 at the boundaries of the range. In spring and summer, most males travel from the ends of the range (Santa Cruz in the north, Pismo Beach in the south) the 60–100 kilometres (40–60 miles) to the centre where some set up breeding territories in the waters populated by females. Some return to the same territory each year, while a few males near Monterey Bay remain in their territories all year round.

A male Californian Sea Otter's breeding territory averages 40 hectares (99 acres), extending 1.5 kilometres (0.9 miles) offshore. Males feed and mate within their territories, but may feed and pursue females outside them. Females breed at four to five years of age but males, although sexually mature at five years, probably take several more years to win a territory. Females occupy about 80 hectares (198 acres) feeding throughout many male territories, and may consistently rest in one but mate with a male from another. Their 'courtship' is a short brutal affair. The male swims up behind the female, clamps his jaws around her snout and the two spin round and round. After between one and four days in her close company, the male departs, leaving the female with a nosebleed as a memento of the event. Females may mate with several males and then, after a brief delay before implantation, start a four- to six-month month pregnancy which culminates in the birth of a single pup. The pup is fed and groomed by its mother until it becomes independent at between five and eight months.

During the past 200 years the Sea Otter has been pushed to the verge of extinction by humans but may now be recovering. It evolved in the shallow waters of the North Pacific one million years ago. Its first contacts with humans were probably occasional skirmishes with Aleut people in Alaska over the past 8000 years. Nonetheless, Sea Otters flourished until the mid-eighteenth century when there were some 200,000 living along Pacific shores from northern Japan to Mexico. Then, in 1741, a Russian expedition led by the Danish explorer Vitas Bering discovered them in Alaska. Bering's ship, the *St Peter*, was wrecked on the Commander Islands and he died, but a handful of his crew survived to carry Sea Otter furs to Russia. Soon this new commodity was worth more than the highly prized coat of Sable, another mustelid imperilled in northern Eurasia by the value of its fur. In 1857 the Russians sold Alaska to the USA for $7,200,000, and Sea Otter pelts repaid the cost within 40 years. By 1911 fewer than 2000 Sea Otters remained worldwide.

Sea Otters were scarcely seen along the Californian coast from 1911 until 1938. After this, miraculously, they increased in numbers, spreading along the coast at 1.4 kilometres (0.9 miles) annually to the north, and twice as fast to the south. Then in 1972–73, the otters faltered at two sandy bays: Monterey bay in the north and Estero bay in the south. Their prey inhabited the kelp forests along the coast and sandy bays were, for Sea Otters, barren desert. Eventually, they leapt the gaps, in 1977 in the south and 1982 in the north.

(Overleaf) Female Sea Otters off the Californian coast may share their territories with daughters and granddaughters. Competition within each group may be reduced by the fact that individuals develop different feeding specialities

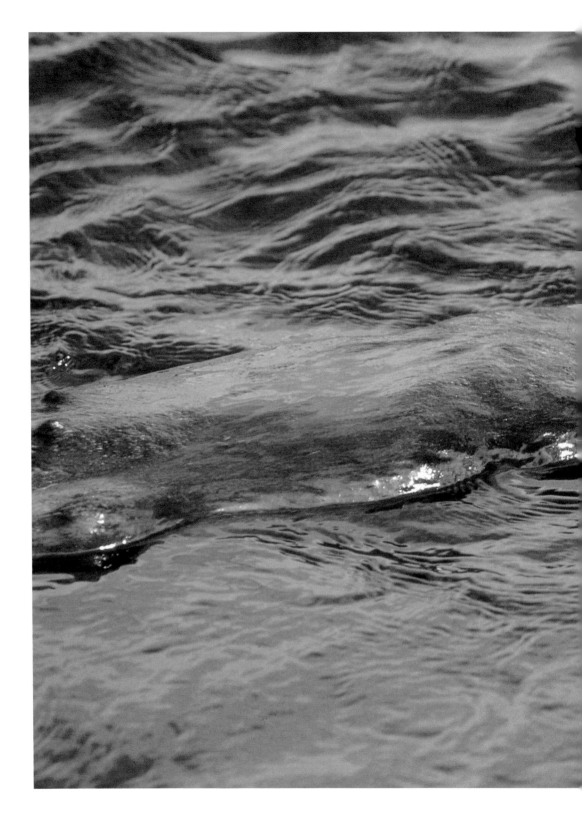

As they recovered, the Californian populations became the focus of controversy. Sea Otters have big appetites and take abalone and crab, which are in demand by humans. Abalone were already being overfished by people but by 1964 Sea Otters were said to be devastating abalone beds along 30 kilometres (19 miles) of California's coastline, and during the 1970s they apparently wiped out a multi-million-dollar abalone harvest at Morro Bay. Shellfish numbers had probably only grown sufficiently to sustain an economic fishery because of the temporary absence of Sea Otters. However, the revival of predators meant that people were faced with the prospect of sacrificing part of their shell-fishing industry or killing some of the most cuddly-looking animals on Earth. Then, the Californian Sea Otter populations stopped growing, perhaps due to deaths in fishing nets, with about 1400 along 350 kilometres (220 miles) of coast. This was a relief for abalone fishermen, but appalled conservationists who feared they could be annihilated by one disaster. Fears were intensified in 1989 when about 3000 Alaskan Sea Otters died in Prince William Sound following the *Exxon Valdez* oil spill. Attempts to capture and rehabilitate oiled survivors cost US$18.3 million and may have saved 222 individuals at a cost of $80,000 each.

Between 1987 and 1990 some 127 Sea Otters were translocated to San Nicolas Island, 130 kilometres (80 miles) south of Los Angeles. Local abalone fishermen were unenthusiastic, but other people may benefit. As well as taking abalone and crabs, Sea Otters also feed on sea urchins which graze the kelp. If uncontrolled by otters, sea urchin populations can explode and destroy large areas of kelp. This brown seaweed's 80-metre (260-foot) long fronds are harvested for food, iodine, potash and medicines. In addition, by providing a habitat for many marine animals, the kelp forest greatly increases the productivity of the Californian coast, facilitating hobbies from bird watching to sport fishing. So, the local economy may be better off with Sea Otters and more kelp, than with no Sea Otters and more sea urchins. However, by 1992 only 15 or so of the translocated otters remained, many having found their way home.

By 1982 the Sea Otter population around the Aleutian Islands in Alaska had recovered to some 150,000 animals. As their numbers grew, Aleutian otters had to spend far more time foraging and gradually changed their diet. Some populations now take a fish-shellfish mix which can support more Sea Otters than shellfish alone. Around Kamchatka Island the Sea Otters have individual preferences, some eating mostly fish and others concentrating on molluscs and sea urchins. These preferences result in fish-eaters tending to have more parasites and white bones, while the bones of shellfish-eaters are stained purple by a pigment from sea urchins. Whether their different diets affect their social lives remains to be seen.

The requirements of size for warmth and strength for digging favoured the development of some hefty otters and badgers. The descendants of *Miomephitis*, a mustelid that lived in Germany 22 million years ago, followed a different route.

Members of the subfamily Mephitinae, or skunks, became much smaller, the 13 modern species weighing between 0.5 and 3 kilograms (1 and 6.5 pounds). Although they remained in Eurasia until between five and ten million years ago when one, *Pliogale*, crossed to North America, skunks are now confined to the New World.

In America, skunks may fill the niches occupied in the Old World by small badgers. Indeed, like miniature badgers, they use their formidable front claws for rooting out food. However, they are much more carnivorous, primarily consuming small mammals and insects, with occasional fruit. Skunks are chiefly nocturnal, and during periods of bad weather in winter will withdraw for prolonged snoozes. Females occupy home ranges of 1 to 2 square kilometres (0.4 to 0.8 square miles), each overlapping to varying degrees with those of other females. They rear their young alone, the males taking no interest in their pups. Each male occupies a territory that encompasses those of several females, but excludes other males.

Small Carnivores face the dilemma of being both predator and prey, and many are therefore discreetly camouflaged. However, skunks all flaunt a natty black and white outfit, which appears to maximise contrast against any background and in any light. The pattern varies between, and even within, species. The flamboyant design warns of their habit of spraying sulphurous-smelling fluid, which can cause temporary blindness, in the face of attackers. Predators rapidly learn to associate the warning colours with an unpleasant memory of the vile, skin-singeing spray.

When threatened, skunks fluff up their tails, stamp their forefeet and walk stiff-legged, showing off their domino livery. If the threat persists, an indignant Spotted Skunk will turn its rear-end towards the predator and whirl up into a handstand, wafting its nether regions in the face of its foe. As a last resort, the skunk drops to all fours and squirts musk from its anal glands in a jet that can travel 5 metres (16 feet). It can aim accurately at the intruder's face because it is equipped with what amount to built-in water-pistols. With the odd exception of bears almost all Carnivores have a flask-shaped glandular pocket on either side of the anus. These are particularly large in skunks, and the secretions copious and potent. Each flask is surrounded by muscles and connected to the outside world by a muscular tube which ends in a nozzle, housed in a little sheath. The two sheaths protrude like guns from a turret on either side of the skunk's backside. Little muscles aim the nozzles and those coating the flask clench tight to fire the jet. Even the dullest of domestic dogs should get the message after only one encounter. Skunks have very few predators.

While skunks are generally nocturnal ground foragers, members of a fourth subfamily, the Mustelinae, found diverse ways to capitalise on being long and thin. Today they number 32 species and, although the distinction is fuzzy, 30 fit into one of four lifestyles: the arboreal martens, the aquatic mink, the prairie polecats and ferrets, and the tundra weasels. All these predators have exaggerated the tubular

The Striped Skunk's minstrel colours may serve as an aide mémoire *to predators of their noxious chemical defences*

design of their ancestors to follow prey into tight corners. Two others, the Wolverine and Ratel, in contrast, have specialised as larger predators in harsh landscapes.

The Wolverine is reminiscent of *Aelurocyon*, the early American digger (Wolverines probably belong to the weasel side of the family, although one school of thought is that they are related to badgers). At 15–25 kilograms (33–55 pounds), Wolverines are the biggest terrestrial mustelids. They have a circumpolar range in cold northern forest and tundra, where their large, fur-covered feet make perfect snowshoes. Scandinavian Wolverines tend to hunt, and can kill winter-weakened Reindeer 10 times their own weight, springing on their backs to overpower them. In other areas, such as Montana, they are essentially scavengers. Wolverines can smell carcasses from more than 3 kilometres (2 miles) away, and their strong jaws splinter frozen meat and large bones. Their habit of raiding trap lines has made them a target for trappers and given them a bad name: Eskimos call them *Kee-wa-har-kess*, the Evil One, and early American settlers named them Glutton or Devil Bears.

In Montana, Wolverines cover vast winter areas of about 400 square kilometres

(150 square miles), while in Alaska, where winter food is even more sparse, their ranges are almost 700 square kilometres (270 square miles). Sometimes they make long excursions of over 160 kilometres (100 miles), travelling away from home for 30 days. They roam high ground at around 2000 metres (6500 feet) in summer, and descend to 1300 metres (4300 feet) in winter. Their summer diet is diverse, including mammals and fruit, but in winter they live mainly on carrion or caches made up to six months before. Montana Wolverines live much like miniature bears, with which they compete for carrion. However, Montana Wolverines are not territorial, perhaps because their social lives are destabilised by the high death rate inflicted by trappers. Elsewhere, female Wolverines maintain territories encompassed by the overlapping and larger ranges of several males.

The 12-kilogram (26-pound) Ratel is the Wolverine's opposite number in the arid south, from Turkestan to the Cape. Although shunning its own kind, the Ratel or Honey Badger has formed an astonishing alliance with a bird called the Honey Guide, which shares its taste for honeycombs and bees' grubs. The Honey Guide can presumably break into some bees' nests unaided, but has no equipment for breaching the defences of the biggest and best nests. So, having located such a nest, it sets off in search of a Ratel (or a human hunter-gatherer). When the Honey Guide (aptly known to science as *Indicator indicator*) finds a Ratel, it flits from branch to branch in the direction of the hive, flashing its tail alluringly. The Ratel responds with grunts, which presumably betoken good will, and follows its feathered friend. At the nest the bird falls silent as the Ratel searches for, and then demolishes, the nest. The hive-opener makes no special provision for the guide, but there are always enough grubs and wax left over to repay the bird.

Nobody knows whether the Ratel's liaison with the Honey Guide is learned or innate. Perhaps the Ratel comes to associate the bird with the location of bees' nests, and the bird associates the Ratel with opened nests. The special fluttering movements that the Honey Guide uses to lead the Ratel may originate in the mobbing behaviour used by some small birds to bombard marauding Carnivores. An ancestral Honey Guide fluttering antagonistically over a Ratel might have flitted on to a bees' nest, thereby launching the partnership. Ratels may also be followed by an entourage of Chanting Goshawks and Black-backed Jackals, eager to pilfer prey.

Among the smaller mustelines, the martens persist in the ancestral trade of hunting in tree tops, a lifestyle extending back to the earliest Carnivores almost 50 million years ago. Today, the most acrobatic is the Pine Marten which can outpace a squirrel through the canopy, although it hunts mainly on the ground. Like most martens the Yellow-throated Marten of Russia occasionally takes eggs, and one

(Overleaf) The Wolverine is the largest member of the musteline lineage and can kill reindeer ten times its own weight

captive specimen was seen to crack them by clasping them in her forepaws, rearing up and throwing them to the ground. With a canine tooth she then made a hole just large enough to lap the contents without spilling any.

The largest marten is the Fisher, with 6-kilogram (13-pound) males twice the weight of females. In the hardwood-hemlock forests of Upper Michigan, Fishers divide their time between the ground and trees, where they are more agile than cats but less so than Pine Martens. They sometimes roam haphazardly in forests seeking hares, and at other times they go Porcupine hunting, following straight routes of up to 6 kilometres (3.7 miles) between all the Porcupine nests they know. A female Fisher can kill a Porcupine nearly four times her own weight. Faced with a Fisher, a Porcupine tucks its face against a tree trunk, and every so often charges backwards, tail flailing, in the hope of colliding with the predator. Being low-slung, Fishers can attack Porcupines on their own level, aiming at the face. This puts them at a great advantage over the lofty Lynx and Coyotes, which find themselves staring down on the Porcupine's spines. So successful are the Fishers that they appear to regulate populations of Porcupines, which are thought of (perhaps unfairly) as destructive pests of forestry. Unfortunately, people realised this only after they had exterminated Fishers in many areas. In 1962 Fishers were reintroduced in Michigan and Wisconsin: 13 years later there was one Fisher every 512 hectares (1265 acres) of forest and Porcupines had declined by 76 per cent.

The 4–6-kilogram (9–13-pound) Tayra from the forests of Central and Southern America is arboreal, and preys on birds and small mammals. However, it differs radically from other mustelines in also being partial to fruit. In addition, while the rest of the subfamily lead fairly solitary lives, Tayra are commonly seen in pairs and sometimes in larger groups, probably extended families.

The same body that allows martens to squeeze into nests and tree-holes as nimbly as the prey they chase could be adapted to hunting nooks and crannies on the ground. The first Carnivore known to pursue prey underground was *Zodiolestes*, whose Greek name aptly translates as 'robber of animals'. It was a sinuous cat-sized animal adapted to burrow-hunting. Its innovative tactic was revealed by the fossilised evidence of a misfortune that occurred 24 million years ago. A *Zodiolestes* had followed prey down a long thin corkscrew-shaped shaft, excavated 2 metres (6.5 feet) vertically into the soil, leading to a labyrinth below. Predator and prey had spiralled down until, just as the *Zodiolestes* lunged forward to deliver the *coup de grâce*, it wedged firm in the burrow and died. Its remains were turned to stone, immortalising the moment, but no clue to the reason for its demise was sculpted on the fossil. Perhaps the prey delivered a lethal defensive bite, or perhaps the predator was just too old and stiff to continue down the spiral squeeze.

This first known burrow-hunter was from a now extinct mustelid subfamily, the cat-like oligobunines, and its prey were hefty, burrow-dwelling beavers called *Palaeocastor*. Its beaver prey lived in large colonies, so *Zodiolestes* thrived by specialising on them until the prey died out taking the predator with them. This scenario has a mournful resonance in the contemporary story of the Black-footed Ferret, the only representative of the polecats in North America. The Black-footed Ferret specialised in hunting the North American ground squirrels called prairie dogs. It lived, hunted, bred and died in the confines of prairie dogs' colonies, and its subterranean, nocturnal habits have always kept it out of the public eye. Indeed, when the famous ornithologist, John James Audubon, illustrated one he was accused of inventing the species to embellish his work. A century ago there were 5000 million prairie dogs in America and Black-footed Ferrets probably flourished. Since then their prey have been decimated by poisoning campaigns, the ferret is extinct in the wild, and was reduced to 10 individuals in 1986. In 1991 it survived in only three captive colonies totalling about 180.

Other prairie burrow-hunters include the Steppe Polecat of Eurasia, ancestor to the Domestic Ferret, the Marbled Polecat of south-east Europe and Arabia, the Grison of the Argentinian pampas, a hunter of guinea pigs, and the African Polecat or Zorilla. Some of these mustelines illustrate a remarkable convergence with skunks, despite over 20 million years separation between the subfamilies. For example, the 1.4-kilogram (3-pound) Zorilla has skunk-like black-and-white markings. When threatened, it faces away from the adversary, fluffs its tail and erects a crest along its back to show its stripes to best advantage. If pressed it squirts its anal sacs, swinging its haunches in an arc to spray the enemy. If this fails to deter the aggressor, the Zorilla shams death.

Zorillas occur alongside the 300-gram (0.7-pound) African Striped Weasel which can squirt anal scent up to 1 metre (3.3 feet) when threatened. The two species look somewhat similar, although they are only distant cousins. Their dramatic coloration presumably warns predators of their pungency, and their similarity in appearance may help them divide the task of educating local predators in the meaning of their shared signal. If so, punier Striped Weasel would seem to have most to gain.

A selection of other mustelids have dramatic markings. Some have horizontal head bands, like the South American Grison and Patagonian Weasel. These longitudinal stripes are elaborated on the faces of the American, Eurasian and Hog Badgers. Eurasian Badgers at least are tasty, but Stink Badgers, also longitudinally striped, are unpalatable. Like Eurasian and American Badgers, the Wolverine is immensely powerful, but its side stripe is dimly shaded whereas the Ratel's is strident. Ratels are named after the sound they make just before squirting a foul fluid whose potency deters attacking hyaenas and Lions. The European Polecats, Black-footed Ferrets and Marbled Polecats have white ringed faces. Similar rings appear only faintly on the faces of Oriental Ferret Badgers and all these species have potent anal pistols.

This scattering of strident markings among mustelids is puzzling because the plain otters are most closely related to skunks, whereas the badgers and mustelines

The Zorilla belongs to the musteline lineage, but its markings are uncannily similar to those of skunks

are sister groups. It seems that strident markings have evolved more than once and may warn of either scent or muscle power. No single factor, such as insect-eating or nocturnality, unites all the stripy species. Being small and vulnerable does not lead inevitably to minstrel-markings, as proved by mink, martens and many of the weasels, while being ferocious may not lead to facial stripes as proved by Wolverines. Some may mimic larger cousins, but some may have been adapted to other functions, such as signalling to rivals or mates, or providing a target during greetings.

The northern weasels have opted for discretion and are usually a plain chestnut colour. The smallest of the mustelids, they are the product of a process that began five to seven million years ago with the climatic changes that revolutionised the histories of both cat and dog families (see pages 55 and 121). The northern forests were replaced by tundra grassland, which prompted an explosive evolution of small, burrowing rodents. The tubular mustelids were the right shape to take advantage of the bounty, and simply had to get smaller. The miniaturisation occurred in parallel in two parts of the world. In North America, the Long-tailed Weasel arose just over two million years ago, and shortly afterwards an ancestral Stoat emerged as its mirror image in Eurasia.

During the Ice Age, these diminutive predators found their prey spending much of the year under a blanket of snow. The small size and long body that enabled the Stoat and Long-tail to hunt down burrows also enabled them to operate beneath the snow. They remained separated until half a million years ago when water locked in polar ice caps caused the sea level to fall sufficiently to expose the Bering landbridge again. The Stoat soon crossed to North America, where its fossils are now found alongside those of Long-tailed Weasels in Arkansas. However, its American relative did not venture into Eurasia. Subsequently, the world warmed again, the ice caps retreated, and the Stoat followed the new tundra grasslands north. At roughly the same time, the Stoat's lineage in Eurasia gave rise to the Weasel, which can weigh as little as 30 grams (about 1 ounce). During another glaciation 200,000 years ago, the Weasel also crossed to North America.

The result of these migrations is that today three small weasels live in North America and two in Europe. To distinguish them from relatives also known as weasels, these three might be called the snow weasels. Generally the Long-tail is the largest, followed by the Stoat and then the Weasel. However, the three vary in size across both continents. In the north of America, where the Long-tail is absent, Stoats grow larger; in Spain, where Stoats are absent, Weasels are Stoat-sized; and where Weasels are absent, Stoats are smaller. It seems, therefore, that the variation in size of each species is due partly to the competitive scrum. This is most clearly seen in the diameters of the killing canine teeth which neatly separate the snow weasel species (and within them, the sexes) living in each district. Where the species mix differs, the canine diameters reshuffle accordingly. Other mustelids, such as American Mink, also fit into this sequence in which separation of canine sizes presumably reflects separation of prey sizes. But climate, and thus latitude, may also affect mustelid size. Just as Ratels get bigger towards the south in the southern hemisphere, Stoats generally get bigger towards the north of North America, whereas in Europe the Weasel gets smaller towards the north. As a result, the Weasels in one place may be larger than the Stoats in another.

(Overleaf) The Weasel is the smallest Carnivore and hunts for rodents, like this Yellow-necked Mouse, by day and sometimes sleeps off the meal in the victim's nest

The predatory finesse of these tiny Carnivores is unrivalled. They deliver a lethal strike too fast for the human eye to follow, and a Weasel can kill a rabbit that is five to ten times its own weight. This is equivalent to a male Lion killing a female Asiatic Elephant. The Carnivores that came closest to the Weasels' predatory accomplishments were the sabre-toothed cats that took rhino-sized prey. The snow weasels' prowess arises because of, rather than in spite of, their size. While their teeth can penetrate the thin skull of a rabbit, a big cat's teeth, though proportionally larger, are built of the same material and cannot stand the greater force needed to push them through the skull of a rhino. Furthermore, a Weasel can lift ten times its own weight, but a lion-sized weasel could not. This is because muscle power increases in proportion to the *area* of muscle block, whilst weight increases in proportion to *volume*. Hence to approach the relative strength of a dainty small Carnivore, a large Carnivore must be heavily built with greatly increased muscle area. To lift ten times its own weight, though, a lion-sized creature would have to have a disproportionately, and absurdly, huge area of muscle.

The hunting skills of snow weasels are legendary, and none more so than their habit of entrancing victims with a dance. Stoats reputedly throw themselves into gymnastic convulsions which seem to mesmerise their rabbit prey, allowing the predator to get within reach. Apparently the victims often die from fright. Stoats quite often dance with no audience in sight, and these performances may be attempts to attract prey, or may have their origins in play. Another possibility is that such Stoats are victims of skrjabingylosis – a condition brought about by parasitic worms burrowing into the skull. Skrjabingylosis is probably agonising, so the convulsing Stoat may be demented by pain.

The hunting prowess of the miniature mustelines do not fully protect them from the hazards of their profession. When Long-tailed Weasels were watched hunting Cottontailed Rabbits in an enclosure, a weasel that fluffed its attack had a good chance of being killed by the rabbit. In addition, their small size makes snow weasels prime targets for larger predators. In their deep brown coats they are camouflaged in summer but would stand out like beacons on the snow. So in snowy regions they don a white winter coat. In the case of Weasels, there are two subspecies in Europe: the northerly Least Weasel, *Mustela nivalis nivalis*, and the more southerly Common Weasel, *M. n. vulgaris*. They meet along a 100-kilometre (60-mile) wide line stretching across Sweden at about 60 degrees north, roughly on a level with Stockholm and Uppsala. To the south, Common Weasels stay brown throughout the year; to the north Least Weasels turn pure white in winter; and along the line some are turncoats while others are not.

In contrast, whether Stoats living on a climatic borderline turn white in winter is controlled by a temperature-sensitive gene. The critical temperature for turning white varies from one place to another, and also from one part of the Stoat's body to another. For example, while the flanks and rump turn white at 2°C (36°F), the head and back may change at −1°C (30°F). The result is that Stoats change to the white form, Ermine, where more than 2 centimetres (about 1 inch) of snow lie for at least 40 days annually. Where the climate hovers on the edge of this threshold, piebald Stoats are common. Outside borderline areas, the Stoat is much less temperature-sensitive. For example, an individual transported from the Swiss Alps (a white zone) to England (where Stoats remain brown) stubbornly turned white like its alpine fellows. The Ermine flaunts a black-tipped tail which seems to be a flaw in its otherwise pure white camouflage, but the black spot does have a function. Birds of prey that detect the animal direct their attack at the conspicuous tail tip, and then falter at the last minute, distracted from the cryptically coloured and vulnerable body. Presumably the advantage of putting the predator off its strike outweighs any added risk due to the tail being spotted in the first place.

Another problem caused by their small size is that snow weasels lose heat very fast. Furthermore, their long, thin bodies have a greater surface area and so lose heat three times faster than a similarly sized, more spherical body, such as that of a vole. They cannot put on winter fat without spoiling their trim physiques. Consequently snow weasels struggle to eat enough to balance the energy they burn to fuel a racing metabolism. When running underneath the snow they are protected from the chilling night sky as an Eskimo is in an igloo. When resting they need a warm place and will commandeer the fur-lined nests of the rodents they kill.

The burrowing rodents to which the snow weasels owe their existence proliferated in the Ice Age. Like contemporary lemmings and voles of the northern tundras, these ancient rodents were probably prone to massive fluctuations in population. For example, Brown Lemmings on the Alaskan tundra peak at over 200 per hectare (83 per acre) every four years, and plummet to less than one per hectare (0.4 per acre) in between. So early snow weasels would have been part of a boom and bust economy. This would have put a terrific premium on an ability to churn out young fast to capitalise on a glut of prey. Weasels and Stoats have developed very different ways of doing this, principally because of the difference in their size.

All else being equal, smaller mammals breed faster than larger ones. This is because the metabolisms of small mammals race to make up for the punishing heat loss they suffer due to their size. Everything goes faster in a racing metabolism, including growing up, pregnancy and life. A female Least Weasel's heart beats 500 times a minute and she is sexually mature at three months. Having been born in spring, she may have her first litter that summer, at much the same time as her mother is giving birth to her second litter of the year. Weasel litters average six pups so, if she survives long enough, a female has the potential to produce 30 descendants a year: six in her first litter, another six in her second, plus six offspring from each of the three daughters she is likely to have averaged in her first litter. Her sons are unlikely to breed in their first year. With an average life expectancy of less than one year, such profligacy is essential.

Stoats, being bigger, grow more slowly, and live longer, up to 2.5 years. They cannot grow fast enough to breed during the summer they are born, nor can they squeeze two litters into one year. So they must make the most of their one annual litter by producing it early during the spring abundance of prey. They achieve this, as in bears (see page 173), by delaying the implantation of the fertilised eggs. Delayed implantation is so common in mustelids that ancestral Weasels probably did it, but abandoned the habit to accelerate their birth rate. Female Stoats hold the eggs for nine months before triggering a one-month pregnancy and give birth ten months after mating. The delay makes it difficult for Stoats to anticipate the food supply that will prevail when their young are born. Prepared for the best, they produce on average 8–10 eggs, up to a maximum of 19. If the food supply is poor near the time of birth, they will reduce the litter by a mixture of abortion of unborn young and nestling mortality. In the few good years, most of the litter will survive to take advantage of the bounty.

The female Stoat has another trick to give her a racing start. She mates as an unweaned infant in the nest. The mother Stoat comes into oestrus while nursing, and her suckling daughters' reproductive systems may be turned on by the hormones circulating in her milk. The nestling females mate with their mother's current mate, which is highly unlikely to have been their father. Male Stoats that have been observed carrying food to young were probably following up their paedophilia by investment in the infant mothers-to-be of their unborn offspring.

Mating as a nestling combined with delayed implantation doubles the number of litters an average female Stoat can squeeze into her lifetime. The result is that in a poor year a female Stoat may produce more offspring than a Least Weasel, but in a peak year she cannot produce more than about 13 young. Her life expectancy is, however, considerably longer than that of a female Weasel, and this may increase her chance of breeding in a peak year. Their different strategies mean that neither Weasel nor Stoat is likely to gain the upper hand for long if they compete in a fluctuating environment.

There are other mustelids that, like the Weasel, do not delay implantation and some of these are puzzling. For example, martens invariably show a delay, lasting 300 days in the case of the Fisher, while European Polecats do not. Similarly, Eurasian Otters have conventional pregnancies while the American River Otter delays implantation. Northern Eurasian Badgers have a true pregnancy of six weeks and give birth in February. This would require mating in December, a time when the barrenness of winter drives them into a deep lethargy. So they mate early and delay implantation of the fertilised eggs. Unusually, female Eurasian Badgers have at least two oestruses, one or more in spring and one in autumn, and then give birth the following spring. One possible reason for the second mating season is that there may be a change in the identity or status of breeding males in the group between the spring mating and the birth of the cubs. Furthermore, it probably makes it more difficult for one male to monopolise all matings, and litters of mixed patterning may diminish the dangers of infanticide and inbreeding. It would be interesting to know whether, within a badger clan, cubs sired in the autumn have the same fathers as those sired in spring.

Female American Mink have four receptive periods during the spring when they can be stimulated to ovulate by the vigorous act of mating. The periods are separated by six to twelve days and during each, a new wave of eggs may be shed in the womb. Females that mate early postpone implantation for almost two months, and those that mate late delay for only a fortnight so that all give birth in May. During each receptive period the female may be courted by more than one male, and from one ovulation she may produce a litter with more than one father. Even if she is fertilised during her first heat, she continues with subsequent ones. During any one ovulation, the last of a series of males to mate with her secures the largest share of the offspring. Similarly, over all four heats, the males that mate later sire more of the eventual litter. The female's system assumes that the last in a succession of males is likely to be the best. She probably cannot choose with whom she mates, but her reproductive system allows her to wait for a high-quality mate without risking ending up barren by being too choosy at the outset. The system may also tempt a good quality male to stay around in case the eggs he has fertilised are usurped by a later male. Early males attempt to thwart latecomers by monopolising the females through matings that may last three hours, during which they ejaculate several times.

Among most mustelids mating is a somewhat unceremonious act. The male European Polecat, for example, grabs the female by the scruff of her neck and pinions her with a bite that draws blood while he mates. Female American Mink display their sexual experience in a discrete white badge on the neck formed by white hairs growing where the males' teeth scarred their skin. The male Marbled Polecat displays his libido through fur which can vary from unassuming buff to pulsating orange, and grows ever more vivid as the mating season approaches. The male's hue reflects his testosterone level which, in turn, reflects his dominance. When a patchwork orange male meets a yellowish rival, the brighter coloured individual dominates. A yellowish male that wins a territory changes colour over the ensuing weeks, but if he is deposed his orange livery will fade back to buff.

Most females can be easily overpowered because, in mustelids, males tend to be larger than females. Male Wolverines and Zorillas are 1.5 times the weight of females, male Fishers may be twice as heavy and male Least Weasels 2.24 times as heavy. Males of Eurasian Badgers and most otters are only just heavier than females, whereas males are significantly larger among Sea Otters and Eurasian Otters from northern and western Europe, but not southern and eastern. Generally, the difference is least in species in which pairs share a territory. Another general rule is that the differential diminshes from smaller to larger mustelids, although

this may be biased by the fact that all the small species belong to the musteline subfamily.

It has been suggested that the size differential reduces competition between males and females, just as the feeding specialisations of Sea Otters may reduce competition between females sharing a range. A male Ratel in the Kalahari travelled 174 square kilometres (67 square miles) trekking in pursuit of hares, Cape Foxes and snakes, while a female and daughter cohabited in 54 square kilometres (20 square miles) meandering in pursuit of mice, scorpions, lizards and insects. Diminutive burrow-hunters, especially, need a high intake to run their souped-up metabolism and may have minimised strife by evolving different sizes and thus partially different diets. Competition may also explain differences between species. For example, in Britain male and female Stoats and male and female Weasels coexist and their average weights are, in order, 321, 213, 116 and 62 grams (0.7, 0.47, 0.25 and 0.14 pounds). Similarly, the 770-gram (1.7-pound) Zorilla cohabits with the African Striped Weasel at 310 grams (0.68 pounds), while the 870-gram (1.9-pound) American Marten cohabits with the 2–5-kilogram (4.4–11-pound) Fisher. Their size differences separate off rival species like a sieve: small species can use burrows through which their rivals cannot squeeze. However, this did not defuse worry that 1-kilogram (2.2-pound) American Mink, which escaped from fur farms in Europe, would oust the 10-kilogram (22-pound) Eurasian Otters. Otters specialise in fish, crayfish and frogs, which also form part of this mink's more general diet. In Scandinavia the two species have rather different diets in summer, during which the otter concentrates on large fish and the mink on small mammals. In winter they are thrown into stiffer competition as freezing forces both into open water. Mink become scarce in the lakes where Otters abound, but predominate in overgrown waterways where ice inhibits the fishing skills that are the Otter's forte. So it seems that they can coexist as generalist and specialist. The likelihood of the rare European Mink living alongside the invader is, however, much less.

Another explanation for the size differential in the sexes is that the female's task of rearing young confines her close to the den, limiting her home range and thus her body size. Unaided by her mate, a female Stoat has to catch up to four times more prey daily when nursing. If she was the size of a male, she would need yet another extra vole a day. Hunting for that extra vole would keep her away longer from her young, which need her protection and her body heat for their first five to seven weeks. Similarly, when her kits are young, a mother Fisher needs to eat one hare every four days, one porcupine every 22 days or 1 kilogram (2.2 pounds) of carrion every 5 days. By the time her babies are eight weeks old she needs to double this. She is caught in a vicious circle: producing milk requires her to hunt more,

The American Mink is now widespread in Britain, having escaped from fur farms. It probably does little harm to fish stocks, but can cause declines in Water Vole populations and can damage ground nesting seabird colonies

which causes her to burn up more energy so she must consume still more. The Fisher may travel 10 kilometres (6 miles) to make a kill and nursing pups can force her to hunt 21 hours a day, leaving her exhausted. If the female was twice as big, like the male, her bigger appetite would force her to hunt round the clock, leaving cubs untended, and she would still probably fail to make ends meet.

A similar argument is that the increased girth of pregnant females limits their size. The females of burrow hunters have to enter prey burrows during pregnancy, a time when they are hungrier than ever, and so must be 'scaled down' throughout the rest of the year. The body diameter of a pregnant female Weasel is similar to that of a male; when not pregnant, females are 'scale models' of the larger males.

The other side of the puzzle is why males are big, and this may be due to the different problems faced by the sexes. While a female's reproductive success is largely determined by her ability to secure food, a male's is determined by his ability to secure females. Larger males, with greater muscle power, are likely to be able to travel further in search of mates and, having found them, to dominate smaller rivals. Furthermore, to the extent that they have any choice, females are more likely to select big males as mates in the hope that their sexual prowess will be passed to their sons. Hence males, free from the ties of pregnancy and nursing, have the largest bodies their food supply will allow. The fact that the size difference between the sexes is most exaggerated in the smallest weasels may arise because bigger Carnivores generally require disproportionately larger home ranges since they eat larger prey, and larger prey are generally scarcer. Consequently to reduce their territories by the same percentage (by evolving smaller sizes and thus reduced appetites), the female of a small species would have to lose a much greater proportion of her weight than the female of a larger species.

Among most mustelids, the home ranges of fairly solitary males and females overlap, but the details vary enormously between species. In the most carnivorous species, such as Fishers, mink and Stoats, the smaller territories of solitary females are overlain by larger territories occupied by males. Female skunks and American Badgers are somewhat more omnivorous and, although solitary, their ranges overlap. However male skunks occupy territories that encompass those of several females, whereas male American Badgers seem to overlap widely.

These differences affect the extent to which males try to monopolise females. Some try to keep exclusive rights to several females by encompassing their territories; others roam promiscuously and there is every shade of variation between. For example, in some, such as Stoats, American Martens, American Badgers and Weasels, there is a dramatic change with the onset of the breeding season. The most dominant males abandon what was presumably a feeding territory and roam far afield, dominant male Stoats' ranges increasing 50-fold at this time. Even males that maintain exclusive territories, such as mink, are bombarded by intruders when females are on heat. Whether males attempt exclusive territorial polygyny or roam promiscuously must depend partly on how widely the females are spread out, which depends on their food. Wolverines' ranges, like those of bears, are so vast that male Wolverines probably have little option but to roam promiscuously. On a smaller scale, where 600-gram (1.3-pound) female Siberian Weasels in Japan were widely spaced, each male's territory encircled one female's, but where food allowed the females' territories to be closely packed males moved towards promiscuity.

In contrast, some populations of Eurasian Badgers and Otters, and possibly also Tayras, live in territories where the food supply dictates that the smallest area an individual needs will also accommodate several companions. Each worm-eating badger needs several pastures, each Shetland Otter needs sheltered and exposed coastline, and each Tayra may need trees in fruit. All these foods facilitate group-living, but again details of the food supplies contribute to slight differences in societies. The practicalities of fishing force Shetland Otters to operate alone whereas the worm and fruit-eaters can forage in company. The Tayra's tropical fruit is available year-round, but temperate Badgers face a barren winter that forces them towards lethargy huddled in a communal den where their heart rate falls from 55 to 25 beats a minute.

While sharing a den and a territory with females does not stop male Badgers siring cubs in the neighbouring clan, otters appear to be more faithful. Most otters are monogamous, like the omnivorous dogs, rather than polygynous like the carnivorous weasels and cats. Insufficient is known of their prey to explain why this should be so, although the mobility of fish schools may make their whereabouts as unpredictable as the badger's worm patches. If so, this may create conditions where the smallest territory for one otter will also accommodate a pair. Giant Otters need two habitats: they shift between small creeks in the dry season and further afield in the rains. Perhaps encompassing both facilitates cohabitation. Giant Otters have been seen fishing cooperatively with River Dolphin, and in Peru sociable Giant Otter increase their fishing success by cooperating in the hunt, and may also benefit from group defence against Anacondas which they mob on sight. Safety in numbers against sharks may also encourage female Sea Otters in open waters to gather in herds. Female Sea Otters also congregate because they all rest in the most sheltered kelp beds. Their society, in which males hold mating territories in areas where females congregate, is reminiscent of that of browsing antelope. This parallel is strengthened by the males forming bachelor herds outside the breeding season. Perhaps picking shellfish from the kelp is like picking buds from trees.

The societies of many mustelids have had to contend with unnatural forces too, because the family has not generally enjoyed a warm relationship with people. In 1937 Canada sent a present of 50,000 Ermine pelts for the coronation of King George VI in Britain. Game-keepers have decimated European Polecats, the Black-footed Ferret has fallen to cattle ranching and the Eurasian Otter has shrunk from the poisonous tide of agrochemicals. Against this gloom one mustelid, the Stone Marten, is gloriously turning the tables on people.

Its story began towards the end of the Ice Age when Stone Martens evolved in what is now Turkey and Syria. They spread northward, probably following human migrations, and in the Middle Ages crossed the Alps to Germany. Trade in their skins might have led to their extinction but the market collapsed in the early twentieth century. Then came the challenge of urbanisation, and Stone Martens rose to it magnificently. They moved into towns, and a perverse measure of their prosperity is that the numbers shot in Germany soared from 5000 in 1960 to 50,000 in 1990.

Their success has had unfortunate consequences for some people. The trouble began in 1979, in the Swiss town of Winterthur, where car engines were damaged in a rash of vandalism. Following months' of complaints, patrolman Ruedi Muggler was at the end of his tether. So he planted his own car as a lure and lay in wait. Some hours later, he watched in amazement as a family of Stone Martens trotted up to his car and clambered under the bonnet. Once there, the martens headed for the ignition and other rubberised cables, coolant hoses or insulation foam, and started chewing.

At first it was only around Winterthur that Stone Martens amused themselves in this way. Then the habit spread, much like the practice of Blue Tits opening milk bottles. Now thousands of motorists in Switzerland, Austria and Germany wake to find their vehicles 'martenised'. Audi reports that 10,000 of its customers' cars are attacked each year. One night in April 1988, a single Stone Marten rampaged through a car park in Munich, damaging 100 cars. Martens are also graduating to larger targets: they have brought trains to a standstill by gnawing signal wires, and cut through computer and television cables. A particularly ingenious marten linked up two live wires in a power station. Sadly, the experiment proved fatal and the marten could not enjoy the reactions of the 25,000 people it deprived of electricity for two days.

One explanation for the Stone Martens' behaviour is that mothers give birth on engines because they are warm. For the youngsters born in cars, coolant hoses provide the first potential 'food' to play with. However, video cameras placed under bonnets reveal that martens that have never seen a car before also home in on cables unerringly. Another mystery is why martens should become persistent offenders, returning nightly to chew an indigestible hose. Perhaps the taste of rubber or coolant is attractive to them. Certainly, pungent chemicals have failed to deter the marauding martens. Now Mercedes and Audi have designed high-voltage deterrence kits that use the car battery to produce 4000–8000 volts of alternating current. The shocked Marten's hair may stand on end, but otherwise it is unharmed.

Family Tree: **Mustelids**

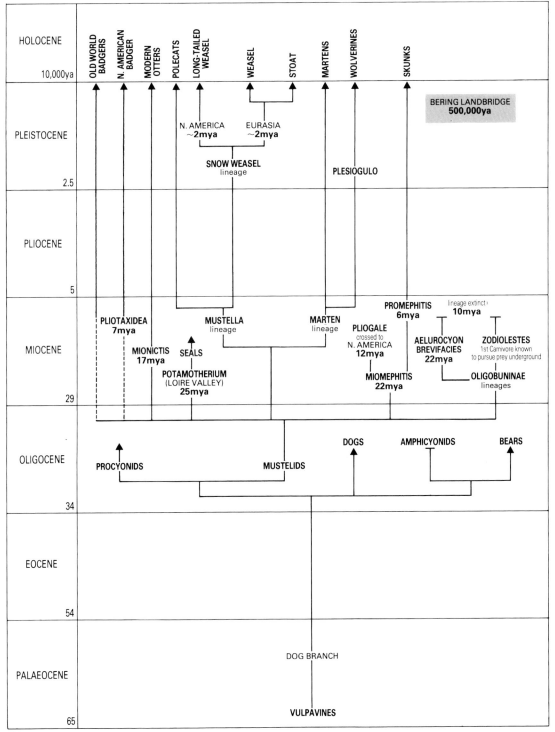

CHAPTER 7

It's Tough At The Top

The dog-branch mustelids were not alone in perfecting the trade of miniature assassin. In the Old World there are cat-branch contenders for the role: the mongooses. Both ventures in miniaturisation were prompted by the same climatic changes that opened up savannas on which small mammals proliferated. However, the cradle of mustelid evolution was further to the north than that of the mongooses and so the mustelids could block advances towards the Bering land-bridge from their more southerly rivals. Further, the insects and reptiles eaten by many mongooses would probably have been unavailable during a winter trek across the northern bridge. Consequently mongooses never made it to the Americas.

The two families of diminutive, long, thin predators are truly rivals, each having produced representatives for various small Carnivore trades. The Marsh Mongoose fills a similar niche to the mink, the Slender Mongoose mirrors the Stoat, and in size but not diet the Dwarf Mongoose is closest to the Least Weasel. But some mongooses evolved a trait that no weasel ever mastered: division of labour in a well coordinated group.

The ancestry of the 36 species of modern mongooses has been sorely debated. They differ from other Carnivores in the bones of their middle ear and in 1916, R.I. Pocock, London Zoo's magnificently productive taxonomist, concluded that mongooses were a family in their own right. Then in 1945, the taxonomist George Gaylord Simpson decided they were merely a type of viverrid. In common with civets and genets, mongooses have teeth much like those of the earliest Carnivores. However, mongooses differ from civets in that most lack a pocket in their outer ear and certain scent glands (see page 27). A mongoose's eyes generally have horizontal pupils whereas, with the irksome exception of the Palm Civet, viverrids have vertical pupils. Furthermore, the chromosomes of all mongooses are remarkably similar, and differ from those of other Carnivores. Such details eventually led to mongooses being separated from the viverrids and placed in the family Herpestidae. Indeed, one modern view is that their closest relatives are hyaenas.

Identifying the earliest mongooses is difficult because of the close similarity between their fossils and those of early viverrids. There are 18–22-million-year-old candidates from Europe, *Semigenetta* and *Leptoplesictis*, and East Africa, *Legetetia*, so either continent may be the family's original home. Early mongooses probably lived like Slender Mongooses, which belong to the genus, *Herpestes*. All members of the genus have teeth that are almost unchanged from early cat-branch ancestors. Slender Mongooses are the most widespread Carnivores of sub-Saharan Africa. These fiery red 600-gram (1.3-pound) Stoat-like hunters prey on small rodents and birds. They can travel rodent runs at lightning speed and are agile enough to take to the trees in pursuit of squirrels. They hunt by day, generally alone or in loose company because their prey is not readily shared. Male and female pairs have been recorded in company with their young, but in other cases two or three males may share amicably a 700-hectare (1700-acre) territory, within which two or three females maintain separate territories. Within these cliques, males are seen equally often with another male or female.

Soon after the mongooses had colonised Africa, Madagascar split off with an early mongoose on board. Its descendants, the Malagasy mongooses, today number six species which seriously confuse the distinction between mongooses and civets. All six have the ear pockets otherwise confined to civets. In addition, two of them, the Ring-tailed and Giant-striped Mongooses, have civet-type scent glands. One possible solution to the enigma was that Malagasy mongooses were not mongooses at all, but viverrids. Anatomists dismiss this idea, but exactly where the six Malagasy Carnivores stand remains a mystery. Only one, the Narrow-striped Mongoose, has been studied in the wild, and it may offer some clues to the origins of mongoose society.

Narrow-striped Mongooses live in Madagascar's dry forests where they forage by day for grubs supplemented, especially in the dry season, by millipedes, snails, amphibians, lizards and mouse lemurs. Throughout the dry season they forage alone or in pairs, sometimes nesting together at night. In the wet season they band together to forage, and reputedly cooperate in pursuit of mouse lemurs and chameleons. Despite this annual fission and fusion, which presumably reflects seasonal changes in their food supply, social ties amongst Narrow-striped Mongooses remain strong.

Over a million years ago, African mongooses travelled east and colonised Asia, where today 12 species flourish. Like the Slender Mongoose, they belong to the genus *Herpestes*. The Small Indian Mongoose is best known, although this 800-gram (1.8-pound) rodent-hunter has only been studied on foreign soil. In an attempt to control rats in sugar-cane plantations, Small Indian Mongooses were shipped to Jamaica in 1872 and to Hawaii and the Fijian island of Viti Levu in 1883. On Viti Levu they live at huge densities with at least 50 per square kilometre (130 per square mile). Their social arrangements are uncertain, but it seems that males occupy rather larger ranges than do females, and that males and females overlap widely as they forage alone by day.

The Ruddy Mongoose of India and Sri Lanka hunts largely by night, alone or in pairs, for rodents and other small vertebrates

In the plantations of Viti Levu, the Small Indians prey heavily on both native and introduced rats. Despite this, they have proved ineffective at controlling the rodents, and their introduction has had unforeseen and unfortunate consequences. They are threatening the existence of several native species, including two rare frogs. They have also become the principal vectors of rabies in Grenada, Puerto Rico, Cuba and the Dominican Republic. By 1973, attempts to control mongoose rabies were costing the government of Grenada 0.4 per cent of its annual revenue. At first mongooses were killed but their populations there still average a staggering 60 per square kilometre (160 per square mile). Now a new tactic is being tried. Unlike Red Foxes, the principal European vector of wildlife rabies, Small Indian Mongooses can recover from the disease and then remain immune. The greater the proportion of immune mongooses, the greater the likelihood that rabies will burn out. In some parts of Grenada almost half the mongooses have become naturally immune, and killing them could simply make way for a new generation of sus-ceptible mongooses. So instead, people are trying to make more mongooses immune through baits laden with anti-rabies vaccine. The fact that mongooses can survive a disease that is fatal to most members of the dog family raises the possibility that rabies first evolved in mongooses, giving their immune systems more time to get the better of it.

Like the Malagasy and Asian mongooses, few of the rest of the 36 species alive today have ever been studied. At least 22 species are said to be solitary, but exactly what this means is seldom explained. To judge by the Slender, Selous and Cape Grey Mongooses, and a handful of others, it is likely that these 'solitary' species live in pairs. Their mixed carnivorous diet is rather reminiscent of New World skunks. Most of them hunt small vertebrates, often by night, and often in forests. All at least supplement their diet with insects, and all enjoy eggs. Marsh Mongooses break eggs by grasping them in their forepaws, standing on their hindlegs and throwing the eggs down. Malagasy mongooses break them by clutching them in all four feet while lying on their sides and then throwing the eggs backwards. The more carnivorous species may eat their cousins; for example, Egyptian Mongooses are reputed to kill Dwarf Mongooses.

Some other species are more insectivorous, and these are nudging towards greater sociality. Among the most beautiful are the 800-gram (1.8-pound) Yellow Mongooses with white tail-tips. They may be seen travelling in pairs as they trot with mincing gait across the arid lands of southern Africa. Pairs of Yellows may share a den with other members of an extended family of 10 or more, and one warren of 50 is on record. They may also cohabit with ground squirrels, both species heeding each other's alarm calls when danger threatens.

In the Serengeti plains, 4-kilogram (9-pound) White-tailed Mongooses forage for insects by night. Females, which are probably members of an extended family, have overlapping ranges and form loose-knit clans, somewhat reminiscent of Red Fox groups. Males in the Serengeti occupy larger ranges which sometimes overlap the territories of more than one clan of females. Elsewhere, White-tails are seen foraging in pairs, or a male's territory may overlap the smaller territories of several solitary females.

Egyptian Mongooses are abundant over much of Africa, and in Israel, where they hunt at dusk and dawn for a mixed diet of insects, reptiles, birds and small mammals. In East Africa they tend to travel in pairs, but in Israel two or three females share a territory with one male. All females in the group may breed, and each will nurse the others' young. The male, which is much heavier than the females, is a diligent father and guards the kits. All the adults give food to the youngsters until they are yearlings. Individual variation makes it hard to generalise about Egyptian Mongoose society. For example, of two males studied in Spain, one spent almost 50 per cent of his time with one female and her young, whereas the other spent 5 per cent of his time with his family. These Spanish Egyptians cooperated by hunting as a family in rabbit warrens – each taking a different entrance to the warren. When the family was attacked by a Pardel Lynx the male fought the Lynx while the female escaped.

The Yellow, White-tailed and Egyptian Mongooses are leaning towards the one trait that radically distinguishes the mongooses from their weasely counterparts: eight of the 36 mongoose species are flamboyantly sociable. They are sociable not merely in the sense of cohabiting Eurasian Badgers (page 182), or even Giant Otters (page 190). Their societies are as wondrously complex as those of any mammal, apart from the apes. They exist as cohesive groups, bound in mutual dependency. These remarkable species include the Banded, Liberian and Gambian Mongooses, the Kusimanse and Angolan Mongooses, the Dwarf Mongoose and the Grey Meerkat or Suricate. All belong to one subfamily, the Mungotinae, so perhaps their compulsive sociability descends from a single ancestor.

The best studied of the gregarious species are the Dwarf and Banded Mongooses and the Grey Meerkat or Suricate. The 1.8-kilogram (4-pound) Banded Mongooses live in packs that average about 15 (but up to 30). In each pack there are several adults of each sex, of which two, generally the oldest, have dominant or alpha status. The alpha female rules the pack, emerging first every morning and deciding upon the group's foraging route. Females probably remain in their mother's pack, whereas breeding males immigrate from elsewhere. Several adults of each sex breed each year. All the litters of one pack are born within a few days of each other, and are nursed communally by up to three mothers. When the group goes foraging up to six babysitters (but generally two females and a male) guard the den. Some males become almost obsessively helpful, always ready to volunteer for community chores.

There are many similarities between Banded Mongooses and the 900-gram (2-pound) Meerkats of the Kalahari. Their groups number 3 to 30, and an alpha pair is identifiable at the hub of each. The alpha female seems to be the principal breeder in most groups, but up to three females may be pregnant simultaneously and, if more than one litter survives, females will nurse each others' young. There are many variations, though, including the case of a non-breeding female that came into milk and acted as a wet nurse to the young of an alpha female. In the 300-gram (10.5-ounce) Dwarf Mongoose's society only the alpha female usually breeds. However again, when a dominant female Dwarf Mongoose died leaving an unweaned litter, a subordinate female came into milk and nursed the orphans successfully.

Exactly when mongooses evolved gregarious bands is unknown, but perhaps it coincided with the opening up of plains seven to five million years ago. This event led to the evolution of vast herds of fleet-footed ungulates which produced copious dung nourishing a massive population of insects. Many of the insects were most

The Marsh or Water Mongoose of Africa is semi-aquatic although, unlike its mustelid counterparts the minks, it has no webbing between its toes

readily caught in the warmth of the day, and day-active, insectivorous creatures, including mongooses, arose to take advantage of this copious food supply.

It seems odd that none of the mongooses' opposite numbers, the mustelids, followed this path. Amongst the weasel-types, many are small and at least partly diurnal, yet none is dedicatedly insectivorous. This may be because the mustelids evolved further north, where the steppes of Eurasia and North America supported herds of Saiga Antelope and Buffalo. Perhaps the cooler winters in the north made insect abundance highly seasonal, and thus precluded the evolution of a full-time insect-eater.

The crucial combination that distinguishes all eight pack-living mongooses is to be both diurnal and insectivorous. Most of the mongooses that hunt alone or in pairs, by night or by day, depend on small rodents, birds and lizards. Such prey is thinly spread and not easily shared. So the carnivorous mongooses can neither benefit from hunting in company, nor can they afford to do so. The insects taken by sociable mongooses, though, are sufficiently abundant to be hunted in company. A group of Meerkats can hunt side by side without getting in each other's way. Furthermore, they can forage in an area on one day, and return the next to find almost as many insects again. In contrast, an imaginary group of carnivorous mongooses that decimated the mice in an area would find the burrows bare for a long time thereafter.

The more mongooses there are in a pack, the faster they will deplete the insects in an area, with the result that large packs travel further each day than do small ones. However, it is not long before the insects are back again, so the cost of companionship is not great. And the advantages of comradeship are enormous to creatures whose small size puts them at continual risk from predators.

There are some insect-eating mongooses that hunt at night, such as the Bush-tailed Mongoose of East Africa and the Selous Mongoose of southern Africa. Oddly, these nocturnal creatures have not been driven to sociality by fear of predators and they forage alone. Perhaps this is because twenty pairs of eyes are just as blind as one to the approach of an owl on a moonless night, and listening and smelling can be done almost as well single-handedly as in a group. Predation at night may anyway be a smaller risk. Another possibility is that there may be something, so far unknown, about their food supply that makes sociality less convenient.

The greatest threats to day-active insectivorous mongooses in open country are the birds of prey that could swoop on them. Their predicament is epitomised by the Grey Meerkats of the Kalahari desert. The Meerkats' group-living has grown out of two compelling but incompatible preoccupations. On the one hand they are rapacious hunters; at every footstep a hole is pawed, a crevice sniffed. When they scent a grub below, they dig feverishly, with heads wedged deep in tunnels and rumps exposed unprotected aloft, amidst a cascade of sand as long claws flail deeper and deeper in pursuit. On the other hand, every Meerkat is obsessed by the need to scan the skies for raptors. Like the Weasel, their dilemma is to be both predator and prey. Unlike the Weasel, their feeding habits have allowed them to crack the problem. Alone, individuals can only trust to luck and waver appre-hensively between feeding and vigilance. As members of a team, they can share the burden of looking out for danger.

Meerkats have developed two different approaches to the problem of watching for predators. In the first, as the group forages, individuals frequently pause, stand up and scan the sky for a few seconds. In general at least one individual is always looking around but, the larger the group, the less often each has to undertake this disruptive chore. Nonetheless, even in the largest groups, each Meerkat's hunting is disturbed every few minutes. In the second approach, one individual takes full responsibility for guarding the group. Among Kalahari Meerkats, the would-be sentinel detaches itself from the frenzied business of foraging and runs to a vantage point. Often it will clamber a metre (3.3 feet) or more up through the fearsome thorns of an acacia bush or to the top of a termite mound. There, it will perch, wavering in the wind, lambasted by the Kalahari sun which can scorch the sand to 60°C (140°F) or more, and exposed as a beacon to predators. Meanwhile, comforted by uneasy peeping sounds that signify a guard is on duty and alert, its companions forage with scarcely a glance skywards. When the incumbent eventually comes down, hot and hungry, another will run to take its place as sentry.

All members of a Kalahari Meerkat mob take a turn at guard duty, except the babies, but each does so to a different extent. The dominant or alpha male only rarely acts as sentinel, and then generally only for 10 minutes or less. In contrast, the senior male lieutenants volunteer often and for lengthy stints. Their devotion to duty can seem obsessive. They seek the highest vantage points, 4 metres (13 feet) or so up, and often stay on duty for an hour or more. Several stints in a day represent a serious loss of time to an animal that must forage almost continuously to fuel its perpetual digging. Adult female Meerkats also guard, but typically for shorter periods and often from unambitious vantage points. Young adults of both sexes guard the least.

Dwarf Mongooses are also bombarded by predators, in some habitats at the nerve-racking rate of 1.5 alarms per hour, of which almost one in ten may escalate to an attack. Almost a fifth of the Dwarf Mongooses' day is spent hiding during air raids. Being very small is a mixed blessing to Dwarfs because although it allows more of them to live together on a given food supply and so increases their capacity for cooperative vigilance, it also increases the number of predators they must be vigilant against. Unlike Meerkats, Dwarf alpha males are particularly active as sentries. Being a sentry is a dangerous job, and the most perilous moment is the

(Overleaf) Scorpions are a prized delicacy of Grey Meerkats in the Kalahari desert. The Meerkats eat these big-pincered Opisthopthalmus *scorpions tail-first, but small pincered, more toxic,* Parabuthus *species are eaten head-first*

changing of the guard. The individual coming off duty often has to run alone for up to 100 metres (330 feet) to catch up with its companions. During these lonely long-distance runs, it is all too susceptible to attack.

With danger lurking everywhere, Dwarf Mongooses need all the help they can get, and this has led to a remarkable mutualism between them and two species of hornbill. These birds are insect-eaters and forage together with the mongooses. The Dwarfs allow this because the hornbills are sharp lookouts. With a hornbill in attendance, the mongooses spend less time scanning the horizon and their system of sentinels may be relaxed. If a hornbill sees a predator it gives a distinctive 'wok-wok-wok' alarm call. It will warn the mongooses even of raptors that pose no threat to itself, but will make no call if the raptor is neither one of its own nor the mongooses' predators. In return, the mongooses inadvertently flush prey which the hornbills would otherwise not find. Other birds join in too, such as the White-crowned Bush Shrike, the Drongo, and the Lilac-breasted Roller.

Each morning hornbills remember which termite mound their Dwarf Mongooses slept in the night before. At sunrise they perch in the trees around the mound, waiting for the mongooses to emerge. If they are kept waiting for more than an hour, the birds will become impatient and fly down to the mound, poke their bills into the ventilation shaft and let out a loud 'wok'. The mongooses respond to this wake-up call by emerging, bleary-eyed. The birds chivvy them into action by flying to the top of the mound, walking down the side of it, and then flying or walking back to the top and repeating the sequence. The mongooses soon take the hint and set out foraging.

Dwarf Mongooses may awaken to find no hornbills if, for example, the birds were disturbed the previous evening and did not see where their mongooses decided to spend the night. On these occasions the mongooses delay their foraging, loitering uneasily around the den for the birds' arrival. They will leave to forage only reluctantly if the birds let them down. As a general rule, the number of attendant birds depends on the number of mongooses in the pack. The hornbills follow the mongooses, foraging in the grass on foot and grabbing any insects or small reptiles that their beaters disturb. Both eat the same prey and often a hornbill will snatch an item of prey while it is being pursued by a mongoose. When the food is restricted to a small patch, such as beetles in a dung pile, the Dwarfs chase off the hornbills. An irritated Dwarf treats the delinquent hornbill like a mongoose, growling and hip-slamming as it would to a rival in the pack. It does not perform the displays, such as lunging, biting and spitting, that are generally used against other species. The hornbills give the mongooses special treatment too. When young Dwarf Mongooses first go foraging they are well within the prey size relished by the hornbills. Although the hornbills may treat the youngsters with a certain lack of respect, hurling them aside in competition for food, they refrain from eating them.

Apart from providing an early warning system, being small and sociable offers

Dwarf Mongooses are the smallest of the herpestid family, and live in the largest social groups. They forage for insects by day, and sleep in termite mounds by night

other important benefits to the eight species of diurnal, insectivorous mongooses. A group of Banded Mongooses may weigh 60 kilograms (132 pounds), as much as a formidable wolf, yet their small size allows them to chase prey down holes through which no wolf could squeeze. The 30-gram (1-ounce) Least Weasel caught alone by a fox can rely on nothing more than its own wits for salvation. Although ten times heavier, the Dwarf Mongoose is still a titbit to many predators, but it is a titbit with up to 20 companions, which together have 880 teeth that can nip the adversary from a whirlwind of different angles.

Coordinated attack on predators is a trait common to all the sociable mongooses. The power of mongoose collectives was first observed when a Martial Eagle, a magnificent bird with imperious crest and huge wingspan, was seen swooping down on a foraging group of Banded Mongooses. The bird grasped a mongoose and flapped heavily upwards to a nearby tree, with its writhing cargo. By the time the eagle had landed in the branches, the bereaved mongooses were hurtling along the ground towards it. On reaching the tree, the Banded mob climbed higher and higher until the dominant male reached the branch on which the eagle perched and lunged at it. The eagle recoiled, fumbled, and dropped its prey. The seemingly doomed mongoose had scarcely hit the ground before it was up and running, back in the throng of its companions, and miraculously unscathed.

When a Dwarf Mongoose is caught, it has another trick to play: at the feel of grasping talons across its back, a Dwarf freezes into poker-stiff paralysis. As soon as the pressure across its shoulders relaxes, the seemingly lifeless mongoose is revitalised and runs for its life. Doubtless 'playing possum' minimises the risk of injury, for the eagle would dig its talons in deeper if the prey squirmed. It also makes it more worthwhile for companions to launch a rescue. In Kenya's Taru desert, almost half the Dwarfs caught by raptors were freed when the pack mounted a counter-attack. On one such occasion the retaliatory mob collided with a Goshawk so hard that it was bowled off its feet.

Meerkats will confront a Black-backed Jackal, their tails poker-stiff and vertical, and fur fluffed out so that their bodies seem to fuse together. They leap in a rocking-horse motion as, with mouths agape and hissing, they edge towards the enemy like an amoeboid mega-mongoose. The lead position in such perilous encounters is taken by the alpha male in Grey Meerkat and Banded Mongoose society, but amongst Dwarf Mongooses it is the lieutenants, the biggest subordinate males, that spearhead the attack. Indeed, a puzzling difference between the species is that the alpha Dwarfs tend to hang back, jumping up and down as they vent their spleen against the predator by savaging nearby plants.

Grey Meerkats in the Kalahari desert fall prey to Martial Eagles, and take revenge on smaller raptors, whether or not a companion needs rescue. At the sight of a Chanting Goshawk, which is a relatively lightweight adversary, a mob of Meerkats will bunch up and race towards it. As the Goshwak flaps into flight, the Meerkats run beneath it, flinging themselves into the air in an attempt to grasp the enemy's undercarriage. Similarly, whilst they will only confront an 11-kilogram (24-pound) jackal if absolutely necessary, they scorn the 4-kilogram (9-pound) Cape Fox, a lovely animal with large delicate ears, an outrageously bushy brush and a dainty, high-stepping gait. Having bolted a Cape Fox from its daytime lair, the Meerkat mob speeds in pursuit, squirrel-sized demons at the heels of an enemy whose tail alone is bigger than a Meerkat's body.

Meerkats encountering a 1-metre (3.3-feet) long Yellow Cobra are similarly mer-ciless, hounding the snake until it slides away. They take appalling risks, dodging back and forth as the cobra strikes in response to nips at its tail. Perhaps the Meerkats' harassment drives the snake from the vicinity and thereby reduces the risk of an unexpected encounter in the dark recesses of a burrow. As Rudyard Kipling noted in *The Jungle Book*, 'Very few mongooses, however wise and old they may be, care to follow a cobra into its hole'. Kipling's Rikki-Tikki-Tavi was an Indian Grey Mongoose and as he attacked Karait the snake 'he knew that all a grown mongoose's business in life was to fight and eat snakes'.

Three species are especially famed for their snake-killing: the Indian Grey, Javan

The Indian Grey Mongoose's ability to kill Cobras is as vivid in reality as in Rudyard Kipling's stories. They have an advantage in that mongooses are very resistant to the poisons of snakes and scorpions

and Egyptian Mongooses. A 40-centimetre (16-inch) long Indian Grey has been recorded killing a 1.9-metre (6.2-feet) cobra, but all three species are fiendishly agile. Their prowess is extolled in the *Mahabharata*, the Indian epic from 1000 BC, in which the mongoose is portrayed as a hero and helper to people. Film of an Indian Grey in action against a cobra reveals that initially the mongoose jumps rapidly from position to position, advancing and withdrawing. The Cobra moves more gradually, striking by swinging downward and extending its body. Eventually, the mongoose moves close, and the tiring snake withdraws and begins to lower its head to strike, but the movement is never finished. Flinging itself forward, the mongoose delivers a killing bite to the back of the snake's head.

Kipling correctly rejected the widely held belief that mongooses eat the herb *Ophiorrhiza mungos* to gain an antidote to snake venom. He portrayed Rikki-Tikki-Tavi's victory as a matter of quickness of eye and foot, with the mongoose's jump outpacing the snake's blow. In fact, snake-fighting mongooses have another advantage: they can withstand more than six-fold the dose of venom that would kill a rabbit. Dwarf Mongooses survive Puff Adder and Spitting Cobra bites. When a Dwarf is sprayed by a Spitting Cobra, other groupmates groom the venom out of their eyes and the victim generally recovers within 15 minutes. Meerkats are also relatively immune to scorpion venom. They frequently reach their forepaws under logs and then leap back as, with a deft motion, they flick out a large scorpion. With lightning-quick pawing and jabbing, the Meerkat tries to knock the scorpion off balance and then dash in to bite its sting, leaving it limp and mangled. Very rarely a Meerkat will suffer a direct hit but, apart from licking the wound, it seems unperturbed by poison that can kill a child.

If a Meerkat does get injured, its membership of a group may enable it to survive. An encounter with an eagle left one Meerkat matriarch limping badly, her head matted with blood and the jaw and eye on one side badly swollen. Only a few days earlier she had given birth to a litter which was still confined to the den. The whole group clustered around the wounded matriarch and shepherded her back to the den. She hobbled in and was not seen again that day. Her companions remained uneasily at the den mouth, intermittently disappearing inside. The next day she was too weakened to stand upright for the dawn ritual of sun-basking. Two males huddled next to her as she shivered, and one remained with her for 35 minutes after the others left to forage. Although a nursing mother needs to feed as much as possible, this invalid remained alone at the den with her babies all day. The next day she was still unable to stand upright, or to dig, but she struggled out to forage in company with the group. For three days her companions, especially two large males, fed her with grubs. She pulled through and, thanks to the care of her team, her young survived too.

Instances of tending the sick have been witnessed in Dwarf Mongooses too. The alpha female seems to be the most diligent nurse, but all the pack will tend an invalid until it recovers or dies. Such care is restricted to ailing adults, since the loss of a sick baby is not nearly so damaging to the group.

Collaboration amongst sociable mongooses extends into other spheres as well. Grey Meerkats dig and pass back sand in tandem to maintain the many hundreds of bolt holes scattered within each group's 350-hectare (860-acre) territory. They usually forage independently, an individual passing too close to a companion getting a firm rebuke. However, Meerkats cooperate when one gets on the trail of sausage-sized geckoes (*Chondrodactylus*) that live in 30-centimetre (1-foot) deep burrows, deeper than the length of a Meerkat's body. To excavate them may take half an hour, but the prize is worth it because two geckoes generally cohabit. Two, or very rarely three, Meerkats will share the arduous work of excavating *Chondrodactylus* burrows. In the end, each is rewarded with a large meal.

All the adults in a Meerkat pack toil for the well-being of youngsters, each taking its turn at the drudgery of babysitting. This involves fasting all day, a major sacrifice for an animal accustomed to eating every few minutes. As with sentinel duty, some individuals are especially diligent babysitters. Once the babies leave the den and begin stumbling along with the group, they are in great danger. If a small falcon swoops, the entire group will rush together, flinging themselves in a protective blanket over the youngsters. At first adults bring food chewed up into gruel back to the babies at the den. When the babies are mobile, each apprentices itself to an adult and follows its tutor like a shadow, begging with an incessant caterwauling squawk. An adult donates the best quality prey to its apprentice, which will defend its forager fiercely against other importuning babies.

When a new litter leaves the den, the previous brood are placed in a terrible quandary. Having been fed occasionally by the adults prior to the new litter's emergence, the adolescent Meerkats find themselves being pestered by the next generation of dependents. They feed the babies, but clearly often have qualms about doing so. They will sniff longingly at the infant's muzzle as it chews the food, seemingly struggling against the urge to steal it back. Indeed, adolescent Meerkats can become quite schizophrenic during this period, successfully begging for food themselves and then morosely passing it on to a nagging baby.

As adolescent Meerkats mature they are faced with another problem. A mated pair is at the centre of all three well-studied mongoose societies and members of a large group may have a long wait before they attain breeding status. Some males try to remedy this by adopting commando tactics. Among Meerkats, a group of roving bachelor males may lay siege to an established group, ousting the incumbent males and stealing their territory and their mates. Members of a neighbouring group may also pose a threat. In one remarkable case, the second most dominant male in a Meerkat group appeared to engineer a territorial fight between his group and their neighbours. When the neighbours started to tire, this male led two younger males, whom he had baby-sat, in an attack. The trio ousted the neighbouring males, kidnapping their females and usurping their territory, and the previously second ranking male became dominant in his new group.

Young female Meerkats may elope en masse, deserting their companions to join bachelor males. The matriarch appears to try to stop them from leaving, presumably because she does not want to lose their labour. Lone females may also immigrate into established groups. In one case, an adult female Meerkat was recorded following a group at a distance of about 100 metres (330 feet) for a fortnight. During this period the interloper was a nervous wreck, having no other Meerkats with which to share the burden of vigilance. Then, the breeding matriarch gave birth, and the next morning the interloper was seen emerging from the maternity den. For the next three days she was left in charge of the babies while mother and group foraged. The interloper scarcely ate, her labours seemingly the price of social acceptance. Indeed, if she emerged from the den while the matriarch was in sight the old female would attack her vassal, driving her back to her duties as nanny. Although she had apparently not been pregnant, the babies repeatedly sucked on the interloper's nipples. Over several weeks, attacks on her lost their savagery, and she became an integrated member of the group.

Interloping females similarly buy their way into Dwarf Mongoose groups through hard work, and some may spend more time alone with the young than any other pack member. In both Meerkat and Dwarf society, these seemingly selfless immigrants may gain more from their devotion than mere acceptance. The unrelated

babies on which the immigrant lavishes care will probably be the baby-sitters of her own young if she successfully inherits the breeding female's throne. Alternatively, the immigrant may be related since neighbouring packs frequently swap members.

The trend towards absolute monarchy is more apparent in Meerkat groups than amongst Bandeds, but culminates in Dwarf Mongoose society. Generally only one pair in each group of six to twelve Dwarfs produces offspring, the others helping by babysitting, suckling the young and, later, bringing them food. The breeding pair are usually the oldest animals in the pack, and dominate totally the other members. Once in power, alpha females, and most alpha males, may reign for many years, one captive pack being dominated by an 18-year-old female. Thus their lifetime reproductive success, the yardstick of evolutionary achievement, can be very high. Their small size, many helpers and the clement climate allow dominant female Dwarfs to churn out regularly two or three litters per year with up to six young in each litter. The breeders also savour all the trappings of leadership, enjoying great popularity and being the favourite grooming partners of all members of the opposite sex.

Every adolescent Dwarf Mongoose aspires to alpha status and so faces a dreadful dilemma. It can opt to stay at home, within the security of family and friends, but this means starting the usually long ascent of the social ladder from the bottom rung. The youngster that stays at home turns its back on an outside world where there may be another pack in which a breeder has just died without an heir. Alternatively, there may be other, footloose emigrants willing to form a liaison in which the youngster might swiftly assume high status that it would take years to achieve at home. Each animal has to weigh the odds and they are stacked rather differently for males and females.

Female Dwarf Mongooses are much more likely to remain with their natal pack than are males: half the females stay at home, whereas 95 per cent of males leave. The result is that most packs tend to become matrilineal, the dominant status passing from mother to daughter. Sometimes a deposed dominant appears to die of a broken heart, although in fact the death is due to stress-induced uraemia (retention of urine). If two females fight to succeed her the entire pack will fall upon them, perhaps as a mechanism for discouraging squabbles within a society where each individual depends on its companions. A new matriarch is generally chosen through a 'grooming tournament' during which the two contenders strive to lick the adversary into submission over the course of several days. The contestants repeatedly swap positions as each attempts to groom the other from above, producing copious saliva. The winner may end up mated to her father, or may chose as a mate a male

Infant Grey Meerkats are tended by baby-sitters who frequently sacrifice an entire day's foraging remaining alone at the den with their charges

with whom she has had a long-standing alliance.

Among males young Dwarfs generally emigrate if their pack includes several older individuals of their sex, and they join packs in which there are fewer such individuals. The risks they run in swapping groups and the hard work they must do to gain entry are then rewarded by fewer rungs to climb on the social ladder. As with the Meerkats, a hell-raising band of bachelors may outnumber and overthrow the resident males in a pack. As many as a quarter of Dwarf alpha males attain their thrones during take-overs. More commonly, wandering males join forces with other homeless emigrants to form new packs, or immigrate into established packs at a subordinate level.

Throughout the year, the level of the sex hormone oestrogen in female Dwarfs reflects their rank. Alpha females have the highest levels, and senior subordinates have more than junior ones. This was discovered by ingenious biologists who used the Dwarf Mongooses' penchant for scent-marking any strange object to persuade wild Dwarfs to urinate on to small rubber pads. Thus they could analyse the sex hormone cycles in wild, undisturbed animals.

All females in a pack come into heat synchronously, but usually only the alpha female builds up sufficient oestrogen to ovulate. As a female approaches the mating season her oestrogen levels rise, stimulating mating. Mating provokes a further release of oestrogen leading to more mating. Dominant Dwarfs at the peak of sexual activity may mate 50 times an hour and their oestrogen levels soar. Eventually a pulse of another hormone (luteinising hormone) is released which causes the eggs to be shed into the oviducts ready for fertilisation. This is called induced ovulation and subordinate female Dwarfs rarely reach that stage. For them, low status is like taking a contraceptive pill, although the effect on higher status ones is weaker and some may occasionally ovulate.

All the males brim equally with sex hormones, except immigrant males which have low concentrations during the days when they are being physically thrashed while striving for acceptance. Higher ranking animals have larger testes and probably, thus, higher sperm counts. So a subordinate mating with a female soon after, or before, the alpha male is likely to produce only a trickle of sperm which must join an egg race with those from the more copious ejaculate of the dominant. Furthermore, with greater sperm-manufacturing capacity, dominants can, and do, mate more often. However, dominant males achieve their monopoly of matings largely by brute force. The alpha male hovers in attendance around the alpha female, and attacks any other males that flirt with her. In the Serengeti the alpha male occasionally also mates with a subordinate female, but in the Taru desert this has never been seen. Subordinate males are left to fight amongst themselves for access to lower ranking females. About 10 per cent of these will get pregnant and if a low-status female does have babies, the dominant female will often kill them. The end result is that about 15 per cent of offspring have subordinate mothers, and

25 per cent have subordinate fathers.

So the Dwarf alpha pair maintain their position of nearly exclusive breeders using different tactics: females manipulate hormones, while males simply fight. The reason for the difference is probably that females have more to lose. A subordinate male does not risk much in keeping his reproductive options open until the very last minute. His sperm can be produced in prodigious quantities at little cost, and the strain on his system of throbbing with testosterone and getting into fights would be well worth while if he managed to sneak even a few matings. Also, there is always a chance that the alpha will get run over by a Lion or that the subordinate's pumped-up masculinity will enable him to overthrow the dominant. However, to keep her options open longer, a low-ranking female would have to invest a great amount of energy in producing babies that are then likely to be slaughtered by the matriarch. In addition, if the subordinate is too run down after the birth to play a full part in raising the closely related alpha's litter, her young kin may suffer and she will lose social advantages. Occasionally a subordinate will slip in one or more of her kits and suckle them along with the dominant's. This may be why it is so important for the female to breed synchronously: subordinates' litters produced when the alpha female is not also nursing are invariably killed. More commonly, subordinates come into milk without getting pregnant and act as wet nurses to the dominant's young. Their contribution increases by over a third the average number of surviving young in Serengeti litters. Because of the close relationship between most wet-nurses and the young, this effectively means that each is producing 0.79 of an offspring of her own. This is a bigger reward than the 0.57 kits per litter reared on average by subordinates that actually give birth.

Mongoose society flourishes on the basis that many hands make light work. By the same token, where the work force is diminished survival becomes a struggle. For example, when a female Meerkat has young she is immediately excused guard duty, and devotes all her efforts to gleaning sufficient food to generate milk for her brood. In small groups, this places a great strain upon other members. In a group of four adults the problem reached crisis proportions when one of the three females had babies. She abdicated from all guard duty, and other group members took on a large share of the babysitting. As a result, almost all of the burden of guard duty devolved upon the only adult male. On one occasion he was on guard continuously for two hours and nine minutes. Numbers can be similarly critical among Dwarf Mongooses. Small groups of Dwarfs simply cannot muster the labour to keep the duty rosta of guards running smoothly and the consequences are dire: where predation pressure is high, groups of fewer than five adults invariably fail to rear their young, and within two years all such short-handed groups are doomed.

Conversely, the larger the pack, the more young can be reared successfully. In the Serengeti, groups of 10 or more Dwarfs were able to increase the number of pups surviving in each litter up to a ceiling of 4.3. This is still a very small litter size in comparison with other Carnivores, such as African Wild Dogs and Dholes, that also live in large packs with a dominant pair assisted by the corporate efforts of many helpers. Of course, mongooses are much smaller and more tubular, but the Least Weasel can squeeze in up to 19 babies without going pop. Perhaps the problem is that the young in larger litters are less well developed at birth, and the semi-nomadic lifestyle of gregarious mongooses demands that the young grow up fast. Having depleted the insects in an area the group needs to move on, following a rota of harvesting and fallowing. While the babies are immobile, the foraging rota may have to be suspended or the infants carried from camp to camp, an appallingly hazardous operation. So baby Dwarfs start travelling with the pack as young as 25 days old – the youngest of any Carnivore – and young Meerkats are almost as precocious. As well as having their foraging rota disturbed, group members are gravely burdened by baby-sitting during the 25 days encamped with the babies and their territorial borders may go undefended. Sometimes a single male Meerkat will detach itself from the group for one or two days and single-handedly undertake a border patrol marking all the scent stations.

In the Taru desert, Dwarfs operate their system of harvesting and fallowing with the precision of a farmer's rotation. Each group has an 80-hectare (200-acre) territory containing about 200 termite mounds, and chooses a different one each night as a base camp. The Dwarfs scent-mark the base camp with secretions from anal and cheek glands. They generally scent the north face of the termite mound so that the prevailing trade winds waft the odour through the mound to provide a homely aroma. As days pass the scent wanes, and probably changes in flavour too, so signalling the interval since that sector of the territory was harvested. This information might help the matriarch chart the route of her pack from day to day. The scent might also repel intruders, although it could equally well guide them to the safest trespassing route.

Moving from sector to sector, a Taru pack tours its territory once every 21 days. This is just less than the duration of its scent marks, as was revealed in an ingenious experiment. Captive Dwarfs were trained to sniff a glass slide tainted with odour, and then to run over to several more slides, each marked with a different scent only one of which matched the cue slide. Having found the matching odour, a Dwarf was allowed to run back for a reward. By performing the trick correctly the mongooses demonstrated that they could distinguish a particular scent from all others. By varying the ages of the scents, the researcher determined their durability. She also discovered Dwarfs could identify individual pack members by the fragrance of their anal glands, but not by that of their cheek glands. However, the amount of cheek secretion pasted on to an object was a clue to the marker's status.

By following a three-week rota, pack members arrive back at a sector just as their scent marks there are fading. This raises the question of whether they time their rota to suit the lifespan of the scent or whether the scent has evolved to suit their rota. Brown Hyaena paste is detectable even by people for at least 30 days, and Banded Mongoose scent lasts for five weeks. Doubtless the Dwarfs too could have developed a more lingering scent if need be. So they probably use a three-week scent because its message is only appropriate for that long.

Territories are defended vigorously against competitors, as was observed in a gang war between Meerkats. The first member of one group to spot the rival gang gave a long peep of consternation and soon the whole mob stood together on their hind legs, peering forward and peeping. Jostling for position, craning their necks and trilling, they jeered at the enemy and then, like a commando of leprechauns, began to leap on the spot. Tails held stiffly over their backs, the gangs faced each other and flung themselves into a war dance. Slowly they edged forward until they charged, the dominant male in the lead. Then, as if morale had snapped, they halted, milled around and regrouped, clasping each other about the shoulders in comradely hugs. They dropped to all fours and dug, not as they do for prey, but with powerful rending strokes that sent plumes of dust up. Perhaps this, too, was part of a signal to the adversaries of the gang's strength. Again they charged and, as invariably happens, the larger mob won. The victors indulged in an explosion of sociality, grooming and hugging each other, and wiping their odorous anal pouches across each other's flanks, impregnating all members with their shared perfumes. The dominant male, especially, fell into a frenzy of scent-marking, dragging his anal pouch in ungainly sweeps across logs, burrow entrances and other gang members. Lurching from side to side he swayed into sprigs of vegetation, smearing them with the glandular region of his cheeks, before launching attacks on little shrubs. Then the whole group ripped up more ground and defecated on the cratered surface.

Relations between neighbouring Dwarfs are no less strained. Again, the larger group almost always wins, although a pack of six defending a breeding den once repelled 16 invaders. In the aftermath of a battle, it is not uncommon for a juvenile to swap sides. The ranges of Banded packs overlap, and larger groups, led by adult males, dominate smaller ones.

In these battles, and in all cooperative aspects of mongoose life, there would seem to be a huge temptation to cheat. An individual might hang back in a skirmish, or fail to volunteer for guard duty. Such shirking would break the basic rule of all mongoose societies – united they stand, divided they fall – and may have underlain an extraordinary episode witnessed in a battle between two mobs of Meerkats. One pack was attacking a band of wandering males which were intent on forming a liaison with some of the local subordinate females. After a protracted running battle, the resident pack cornered the intruders in a burrow and tried to dig them out. It was a long and frenzied onslaught, throughout which one subordinate female hung

(Overleaf) In the desert winter Grey Meerkats lose much foraging time waiting for the rising sun to warm them sufficiently. On cold mornings the line-up faces the sun with their dark undersides fluffed up to absorb heat

back, lying in the shade while her companions battled. Eventually the residents gave up the attempt, at which point the alpha female walked across to the shirker and fell upon her with uninhibited savagery. Soon the remainder of the mob joined in, thrashing the female until she was torn and bleeding. It seemed very like a punishment for dereliction of duty.

Such punishment might be preferable to being ousted from the group because gregarious mongooses have evolved such complex societies that they can no longer manage alone. Their cooperation seems to span every sphere of collaboration found scattered in other Carnivores. They share their food supply and the vigil for danger, join forces to repel intruders and predators, and tend the sick, injured and young. Mated pairs are at the heart of their society and, although the evidence is weak, Slender Mongooses and others seem to tend to monogamy as well. There are even indications of friendly relations and territory-sharing among males. The Slender Mongoose has much in common with a mustelid, the Stoat, from hunting by day to black-tipped tails. Male Slender Mongooses and Stoats are, respectively, about 40 and 60 per cent heavier than females. Also, both are adept mousers although they despatch their prey rather differently. The mousing mongooses all aim at the eyes as they deliver a skull-crunching bite to the head, while the Stoat, like other mustelids, uses a neck bite. Yet despite their similarities, the Stoat is highly polygynous while the Slender Mongoose tends to monogamy or even male coalitions. This is just one of many conundrums that arise when the social lives of animals within and between Carnivore families are compared.

One route to solving such conundrums may be to identify the heirlooms from ancestors. Cat-branch Carnivores are typically more carnivorous than those in the dog branch. In three of the four dog-branch families males of most genera are much bigger than females, whereas they are more equally sized in three of the four cat-branch families. An 'heirloomist' might argue that cat-branch Carnivores carry carnivory and equal sizes as part of the baggage of their history. Similarly, adult males of at least one species in all cat-branch families share territories (e.g. Slender Mongooses, Palm Civets, Brown Hyaenas and Cheetahs) but apart from larger dogs and Eurasian Badgers this practice is rarer amongst dog-branch males. So, some contrasts between the societies of the cat-branch Slender Mongoose and the dog-branch Stoat might descend from longstanding differences in the social organisation of their ancestors. Certainly, two species may find different solutions to a similar problem because they approach it from different directions. However, it is clearly possible for mongooses to adopt stoatish ways: one male Egyptian Mongoose observed in Spain occupied a territory that partly overlapped the ranges of five females.

Another route, often the most productive, is to look for differences in circumstances that have fostered differences in species' societies. In the case of the Slender Mongoose and Stoat, both have long-bladed scissor teeth for slicing up rodents, but the Slender Mongoose's teeth are also adapted to insect-eating. Indeed, its diet of rodents and lizards is mixed with grasshoppers, termites and beetles. Furthermore, Slender Mongooses occasionally feed together on a large carcass. While rodents are usually too sparse to sustain more than a solitary hunter, insects tend to be locally abundant and so, like large carcasses, shareable. A similar comparison can be made between the Stoat and Cacomistle (page 150) which lives much like a Slender Mongoose in tropical America. Cacomistles are rodent-hunting procyonids that also eat lots of insects and even some fruit. The fragmentary evidence hints that they too may incline towards living in pairs although the male is almost 20 per cent heavier than the female.

There seems to be an obvious distinction between Carnivore societies in which a male mates with one female (monogamy) or several (polygyny) and these are associated, respectively, with males being a similar size to females or much bigger. Within these categories there are major variations. For example, a male Tiger defends an exclusive territory encompassing the smaller territories of several females, while male Grizzly Bears abandon territoriality and roam promiscuously. Some Stoats are intermediate, maintaining the Tiger's system for most of the year, but turning to the Bear's one during the mating season. A traditional view is that the distinction between monogamous and polygynous societies hinges on whether the male's help is necessary for rearing the young, whereupon he is trapped into monogamy, or whether he is freed to be polygynous because any cost to abdicating paternal chores is outweighed by the rewards of siring more litters. However, there is no evidence that the young of monogamous Coyotes and Eurasian Otters are more dependent on fatherly investment than those of polygynous Bobcats and American Mink, so it is necessary to look for other factors that launched these species onto their contrasting social trajectories.

Food that can be shared at minimal cost might be a key factor in monogamy because it allows parents to share a territory easily. Food is shareable if it comes in big packets, or in irregular patches or is replenished almost as fast as it is eaten. Examples include the large carcasses that feed Coyotes and Lions, the patches of earthworms that feed European Badgers and Liberian Mongooses, the sheltered coves that feed Arctic Foxes, the fruit trees that feed Tayra and Kinkajou, and the termites that feed Yellow Mongooses and Aardwolves. All these Carnivores can cohabit cheaply. Other food, notably small vertebrate prey such as rodents, cannot be shared and animals that hunted them in the same area would be rivals. Parents depending on such food might need double the size of an individual's territory in order to cohabit. This will be particularly difficult during the breeding season when the female striving to nurse and provision her young seeks to save effort by having the smallest territory that sustains them. The cost to the female of expanding her territory to accommodate a male may outweigh the benefits of his fatherly assist-ance. As a result of their unshareable food supply, these species are launched towards a semi-detached sex life in which the males occupy larger ranges.

The less of his food a male takes from each female's territory, the larger his own territory will have to be. If he takes, say, a third of his food from one female's territory he will need about three female territories to live (or more, because travelling further he will work up a bigger appetite). A male expanding his territory may get access to more females, and therefore he may strive to defend even more space than he needs to eat. To fight, and to travel and hunt efficiently in a bigger area, it will pay him to be as large as his niche allows. Being larger means that these males will need even larger territories to sustain them. The bigger his range, the less he will drain the resources of females that he overlaps, and the less his presence will disadvantage the young he may have sired. In contrast it pays females to stay as small as possible to minimise their appetite and thus the time they have to spend away from their young. In competition for food, the male's larger size might worsen the female's predicament in direct squabbles, but improve it by somewhat separating their diets. On this scenario, the distinction between Carnivores with cohabiting and semi-detached sex-lives arises not from greater or lesser need for paternal care, but from the extent to which their food is shareable.

Having adopted a semi-detached lifestyle, roaming males may not have sufficient certainty of paternity to devote effort to tending young. Territorial semi-detached males also rarely guard or provision young, but their paternal energies, spread between several families, may be better rewarded by repelling infanticidal marauders and food competitors. Of cohabiting species, male Aardwolves, European Badgers, Simien Jackals and Red Foxes are all known to cuckold their neighbours. Perhaps having secured a mating with their 'monogamous' partner, cohabiting males set out to secure whatever polygynous liaisons they can. Thereafter, they return to concentrate their paternal care on the one litter in which they have the highest stake. Male Channel Island Foxes, for example, travel widely during the mating season, but accompany only one female during the cubbing season. Cohabiting males that do not philander include those for whom it is hopeless or prohibitively expensive. The former include puny males among Brown Hyaenas and Domestic Cats, whose best chance is to stay at home and sneak a mating, and occasionally to help their kin. However, lightweight males sometimes have hidden strengths, such as the disproportionately large testes of small individual Wildcats which may allow them to make the most of any rare chance to mate. Males that cannot afford to gallivant include gregarious mongooses because travelling alone is simply too dangerous and, anyway, all receptive females will be surrounded by a group. Even worse, the male mongoose and Lion may return to find that their group has been taken over. By staying at home, however, subordinate male Dwarf Mongooses do manage to sneak some matings.

There are many variations among cohabitants, particularly those that form cohesive groups. Most groups have an approximately equal sex ratio and in some

Banded Mongooses forage by day for insects in large groups which may contain several breeding females. They are similar to Grey Meerkats, but occupy richer habitats and are twice the weight

of these, such as Banded Mongooses, Serengeti Spotted Hyaenas and some Eurasian Badgers, more than one individual of both sexes may routinely breed, whereas in others, such as Grey Wolves, Dwarfs and Meerkats, this seems rarer. Of the remainder, there are some groups with fewer females, such as Wild Dogs, Dholes, Simien Jackals and Brown Hyaenas, and some with fewer males, such as Red Foxes, Arctic Foxes, Raccoons, White-tailed Mongooses, Coatis and Spotted Hyaenas. In male-biased groups, only one female usually breeds and, although little is known of paternity, one male probably gains most matings. In female biased groups, generally all adult females breed, but in some populations of Red and Arctic Foxes only one female does. Only one adult male attends most female-biased groups but amongst Lions and Domestic Cats there are several and all may mate. Some populations of Cheetah, Slender Mongooses and Palm Civets blend cohabiting and semi-detached traits when several males share a larger territory that encompasses the smaller territories of solitary females.

At least five questions are important in determining the pattern adopted: is the nature of cooperation incompatible with motherhood, which sex is more useful as a team member, which sex is most likely to succeed as a philanderer, which sex is most likely to secure breeder status either at home or elsewhere, and what are the genetic relationships involved? A further consideration may be which sex is hardest to control reproductively for the parent of that sex? Societies in which more than one female breeds tend to be those with the most shareable food and periods of super-abundance, such as the autumn fruit and nuts eaten by Eurasian Badgers, sea-bird nesting colonies raided by foxes on the Aleutian islands and migrating herds ambushed by Spotted Hyaenas. Typically, cooperation in such societies involves tasks compatible with motherhood, such as communal suckling as in Domestic Cats, Lions, Coatis and Banded Mongooses. An extreme case of female breeding not affecting the work force is the Eurasian Badger which, apart from sharing territorial defence, does nothing more cooperative than curl up together for winter warmth. In contrast, more than one African Wild Dog female breeding simultaneously might seriously weaken the hunting power of the pack. Each extra breeding female would not only increase the number of pups but also remove one more adult from the work force. Furthermore, when conditions are good, one female Wild Dog can churn out huge litters – the record is 21 pups. Similarly, each additional breeding female Meerkat is one less available for sentinel and baby-sitting duty. In a group of Meerkats containing, say, two males and two females, if both females bred the group would be unable to guard and feed the young adequately. However, if two males mated with one female the litter's survival would be unaffected, and anyway the cost of shared paternity would be diminished if the males were brothers. When conditions are good more female Meerkats breed: the dominant cannot produce a huge litter because the peripatetic lifestyle of Meerkats demands rather precocious young and therefore small litters.

An obvious distinction between cohabiting and semi-detached Carnivores is that among the former males tend to be similar size to females, but there are many intriguing exceptions. Females of cohabiting Spotted Hyaenas and Banded Mongooses are larger than males, as are those of Binturong and Fossa whose societies are unknown. Although males are scarcely larger than females in cohabiters such as Bush Dogs, Meerkats, Eurasian Badgers, Giant Otters and Olingos, male Lions are much bigger. This may have started as an heirloom from ancestors that led semi-detached lives as do most big cats. Males are particularly huge where females are so tightly packed that amassing a harem does not require patrolling a large area. Male Polar Bears find females congregated around seals, and approach them with massive bodies inherited from their bellicose Grizzly ancestor. The system used by a particular male depends on its circumstances, tempered by its heirlooms. Although the sexes are similar sizes in Brown Hyaenas, and vastly different in Grizzly Bears, males of both opt for roaming polygyny because females are so widely dispersed that it is impractical to monopolise several. This spacing arises because bears are huge and hyaena food is sparse, and the lack of a fixed mating season in the Kalahari may also favour continuous roaming by males.

Other factors may also affect the relative size of the sexes. For example, Cheetahs could not vary much in size without jeopardising their high speed. Males and females of most otters are similar sizes and this may be because their food is shareable, allowing a female-sized male to cohabit with the female, or because fighting heat loss to cold water forces both sexes to be as big as possible. Similarly, female gregarious mongooses might need to be as big as males because they depend on collective attacks against predators and neighbours. At least in the case of Eurasian Otters the cold-water argument is weakened because in north-west Europe where the sexes are different sizes, it is the males that have grown larger, the females being the same size as their south-eastern cousins. It is an open question whether the north-western food supply is less shareable.

A less obvious distinction between cohabiting and semi-detached species is their system of ovulation. With females spread out in separate territories, the encompassing male might have difficulty in being in the right place at the right time. He might have trouble finding the receptive female, although scent marking makes this unlikely, and he would certainly be unable to serve and guard all of the females in his territory if they happened to come into heat simultaneously. The female's need to wait for the best male may explain why various species with semi-detached sex lives opt for a reproductive system called induced ovulation, in which females come into oestrus but do not shed eggs into their womb until stimulated to do so by the act of mating. Cheetah, Puma, Jaguar, Domestic Cats, Weasels, Polecats, Mink, Raccoons and the Fossa, all lead semi-detached sex lives, and are all induced ovulators. It is thought, but not proven, that they can prolong their oestrus if they are not mated. Spontaneous ovulators, such as ourselves, shed eggs into their wombs in a cycle that is unaffected by mating. If they do not mate on schedule, they miss the opportunity. Among the Carnivores, spontaneous ovulators include the Grey Wolf, Red Fox, Lion, Spotted Hyaena, Giant Panda and European Badger, all of which have cohabiting social systems, and so male and female are already on site together, often sharing a den.

There are exceptions in both categories. For example, Stoats, with roaming males, ovulate spontaneously, perhaps because males mate with a whole nestful of infant females gathered together. Female Domestic Cats and Common Raccoons are both cohabitors and both are induced ovulators. However, in both cases females may again have to wait for the best male because in groups where many females are receptive simultaneously top males may have their hands full. Dominant male Domestic Cats may roam amongst several female colonies and have their attention stretched between many females at each colony.

Irrespective of whether ovulation is spontaneous or induced, a longer period of receptivity increases the opportunity for a female to accumulate a gang of admirers so that the best may win, and diminishes the chance of a good male leaving her for fear he will be cuckolded by a late-comer. It might seem that females would all gain by being induced ovulators with prolonged oestrus. However, there are disadvantages to both. The prolonged heats of Polecats, Ferrets and Mink that are not mated can lead to fatal infections of their genitals. Furthermore, oestrus can be dangerous for females getting caught up in male fights and because mating can be rough, and anyway it is fantastically disruptive: male Red Foxes scarcely eat for the three days they trail a receptive female, male Lions are exhausted and Meerkats are so distracted from vigilance that they are compelled to mate underground. Anyway, in many cases, having secured the best-quality mate the female then has a vested interest in his survival and that is imperilled the longer the strife continues.

It is not known to what extent the two methods of ovulation are female weapons to manipulate, or solutions to accommodate, male behaviour. However, since related species may have different systems it is not merely an heirloom shared by a whole family. Amongst mustelids, for example, although both are apparently polygynous, Weasels are induced and Spotted Skunks spontaneous. This, and many other questions posed by Carnivores, can only be speculated on and many of the guesses about them may prove wrong. For example, the adornment of a male's penis with little spikes was thought to indicate that its mate was an induced ovulator, the spikes serving to stimulate her, but these embellishments have now been found on some spontaneous-ovulating species. Only a tiny minority of Carnivore species have been studied in detail and vastly more information is required to understand their lives. Sadly, the opportunity to investigate the remainder is receding and may vanish because Carnivores are under increasing threat in the modern world. In particular, those Carnivores that have persisted with carnivory face one inescapable fact: it is tough at the top of the food pyramid. Top Carnivores, ranging from the Least Weasel to the Polar Bear, may seem to have prey at their mercy: in face-to-face combat the vole has no hope against the Weasel, and the Wildebeest none against the Lion. However, ultimately, predators are at the mercy of their prey. Without ground squirrels there would be no Black-footed Ferrets and without Ringed Seals, no Polar Bears.

Variations in prey number may be due to several factors, including, sometimes, the Carnivores themselves. It is a commonly held view that a Carnivore has little long-term effect on its prey's populations, merely removing individuals destined to perish through starvation or disease. However, even Carnivores that confine themselves to doomed prey may affect the size of the prey population. In Canada, Grey Wolves take Moose condemned by starvation and inclement weather. In the absence of wolves, such Moose would survive longer and continue to compete for food with healthier individuals, which might lead to mass famine. So, paradoxically, there are times when wolf predation increases the number of Moose by preventing doomed animals stealing food from companions that might otherwise survive.

In some cases, Carnivores clearly limit their prey. Californian Sea Otters limit the numbers of sea urchins and, in Australia, Red Foxes limit Rock Wallabies, with whom they have been thrown together by human meddling. In places where foxes were killed off, wallaby numbers trebled, the vegetation was over-grazed and the wallabies lost weight. Food and weather also limit prey numbers and probably all three factors often interact. For example, Lynx are the major cause of death of Snowshoe Hares, and they weed out the prey in worst condition. The hares get into bad condition when their food plants bristle with toxins designed to make the browsers ill. These chemical deterrents are expensive to make, so the plants produce them most when the hares are most numerous and browsing the hardest. The weakened hares are thinned out by the Lynx. Then the plants, lulled into a temporary sense of security, cut back on the toxins and the hare populations expand again. In this three-tier system, toxins affect the vulnerability of plants to hares and, thus, the vulnerability of hares to Lynx.

As well as having their populations limited by the supply of their prey, many of the top Carnivores also have characteristics that tend to put animals at risk. One is specialisation, such as that of the Black-footed Ferret which linked its fortunes to the prairie dogs. Generalists, like Raccoons and Red Foxes, can abseil up and down the food pyramid, equally adept as carnivore, insectivore and fruit-eater. Another characteristic of vulnerable animals is rarity and top Carnivores will always be rare in comparison with their prey. This is because at each level of the food pyramid energy is spilled, so there is less to go around at the next level up. Finally, the larger the Carnivore, the more food it needs and the rarer it is likely to be.

These characteristics are exemplified by the Jaguar. Being dedicated meat-eaters, Jaguars are both specialists and at the top of the food chain, while their large size increases their vulnerability. These big cats need a lot of space, males in Belize occupying territories of 30–40 square kilometres (12–15 square miles) and females 12 square kilometres (5 square miles). Furthermore, being big, they specialise on big prey such as peccaries, armadillos and deer, which are themselves under threat in the modern world. The naturally precarious position of the Jaguar has been greatly aggravated by people. Starting in the 1940s, spotted cat coats became increasingly fashionable, and the trend accelerated in 1962 when President John Kennedy's wife, Jacqueline, appeared in a leopard coat. Soon some 15,000 Jaguar skins were reaching the US and European markets each year and a good-quality Jaguar skin could command US$20,000. By 1969 the fur industry had devastated Jaguars and, although greatly protected by 1975 legislation, they are still killed illegally for their skins.

Perhaps the greatest threat facing the Jaguar and similar-sized predators arises from the fact that their once-expansive habitats have been reduced to small fragments. Many of the animals persecuted for their fur or because they tend to compete with humans for livestock have been given refuge within reserves and parks. However, even the largest reserves can support only limited numbers of large, flesh-eating Carnivores, each of which needs a substantial amount of space to ensure sufficient prey. For example, male and female Florida Panthers occupy 301 and 194 square kilometre (116 and 75 square mile) territories, respectively, and eat 50 deer-sized prey annually. There is no single piece of wilderness left in Florida large enough to support a viable population. The 30 surviving members of this subspecies of Puma are calculated to have an 85 per cent chance of becoming extinct within 25 years. As a result of habitat fragmentation the top Carnivores now generally exist only in small populations that are a conservationist's nightmare.

A hazard for all small populations is inbreeding, resulting in the loss of genetic variability. When two animals mate, they both pass on genes to their offspring. Each offspring is likely to receive two different genes for each character, such as fur colour. However, close relatives are genetically similar and when they mate, their offspring might receive two identical genes for a character. This will not be a problem if the genes are healthy. Unfortunately, genes sometimes change or mutate during the process of copying them from parents to offspring. An offspring that inherits a flawed gene from one parent will usually receive a healthy gene from the other, and this will override the bad effects of the mutation. Among close relatives, though, there is a risk that both parents will have the same flawed gene and both will pass it on to some of their offspring, which will then develop the mutation. This is why animal breeders talk of the need to introduce 'new blood' into their stock and try to minimise inbreeding.

The Cheetah appears to be naturally inbred. Until about 10,000 years ago, there were five species of cheetah and they probably occurred worldwide. Then, for unknown reasons, four species became extinct. The modern Cheetah was reduced in number but survived and was then decimated about 100 years ago by human hunters. During these population bottle-necks, a handful of survivors may have inter-bred. This may partly explain why the genes of different Cheetahs are now astoundingly uniform, especially those controlling the immune systems. As a result, if a disease-producing organism cracked the defences of one Cheetah, it could wipe out most of the population. This sequence of events may have occurred in Black-footed Ferrets, which became rare, lost genetic diversity and thereafter were devastated by canine distemper.

By reducing large Carnivores to small populations, humans may be pushing other species through population bottle-necks. For example, Lions in the Serengeti plains have long been numerous, but those in the Ngorongoro crater all descend from just six to fifteen survivors of an outbreak of biting flies in 1962. Similarly, the Gir forest lions of northern India numbered less than 20 in the early part of this century, and now total about 250. Just like Cheetahs, the Lions of Ngorongoro and Gir are much less genetically varied than their Serengeti counterparts. As well as increased vulnerability to disease, this is likely to have other far-reaching effects. The genetically impoverished Cheetahs have 71 per cent of abnormal sperm in comparison with 30–40 per cent for most cats. Four male Florida Panthers had an abnormal sperm count of 94 per cent. The Lions that have survived bottle-necks have less male hormones, less voluminous ejaculate and more abnormal sperm than the Serengeti Lions. Captive Clouded Leopard are also worryingly similar genetically and have a high incidence of crippled sperm. The high juvenile mortality and vulnerability to infection endured by Cheetahs due to inbreeding may be just around the corner for many rare Carnivores.

An even more pressing threat to small populations is a run of bad luck. The habitat of Grizzly Bears has been severely fragmented, and these animals are now restricted to six areas with no corridor between them. This has created six sub-populations numbering less than 1000 with individuals unable to migrate from one to the other. One fire could wipe out a sixth of the population. Equally, if all of the few females in an area happened to rear sons one year, the whole subpopulation could be knocked terminally off-balance. Conservationists have tried to calculate how many bears would be needed to give each subpopulation a fair chance of surviving a run of bad luck. The first estimate was 50 bears. Then people took account of the risk of inbreeding, since a tiny minority of males secure most matings, and the estimate rose to 70 to 90 *breeding* bears. Such calculations gloss over many complexities, but carry a clear message: many reserves are probably too small to allow large Carnivores to hedge their bets against disaster. Similar calculations suggest that African Wild Dog populations of less than 350–500, which includes most of their populations, are likely to go extinct. During 1991 the fragmented populations of African Wild Dogs in the Serengeti-Mara grasslands and the Simien Jackals of Ethiopia's plateaux both chanced to suffer what may prove to be irreparable declines due to rabies.

About 40 (or 17 per cent) of the 236 species of Carnivore are at risk of extinction. When the smaller species of the mongoose, viverrid and mustelid families are excluded, the figure rises to 33 per cent. Almost all the big Carnivores are in trouble, and unless we are prepared to accommodate their needs, they will disappear.

The story of Carnivores stretches back more than 40 million years and has involved the evolution of many strengths. The future of these magnificent animals is uncertain, but some aspects can be guessed at. The massive physical strength of a Lion can drag a carcass it would take ten men to lift. But that strength cannot protect them, nor the Polar Bears, nor any other of these innocent killers, which can now survive only in reserves or wastes beyond the reach of people. The Cheetah, a monument to evolution's aimless engineer, can outpace any mammal since the beginning of time, but cannot evade the consequences of its genetic uniformity. Others display strength of numbers in the concerted action by disparate individuals that unites a team of Meerkats and turns what might have been a rabble of African Wild Dogs into a cooperative unit. There is also strength of character, in the versatility and quickness of wit with which the Stone Marten commandeers motor cars, and Red Foxes survive, without degradation, the process of urbanisation. As humans transform their world, these generalists can play by our rules and win. Above all else, the thrill of the Carnivores lies in their individuality, each shaped by their ancestry, their ecology and their society.

INDEX